AGRICULTURAL LAND IN AN URBAN SOCIETY

Owen J. Furuseth
Department of Geography and Earth Sciences
The University of North Carolina
Charlotte

John T. Pierce
Department of Geography
Simon Fraser University
Burnaby, British Columbia

RESOURCE PUBLICATIONS
IN GEOGRAPHY

Library of Congress Card Number 82-18424
ISBN 0-89291-149-2

Library of Congress Cataloging in Publication Data

Furuseth, Owen J., 1949-
Agricultural land in an urban society.

(Resource publications in geography)
Bibliography: p. 81.
1. Land use, Rural — United States. 2. Urbanization — United States. 3. Conservation of natural resources — United States. 4. Regional planning — United States. 5. Agriculture — United States — Planning. 6. Land use, Rural — Canada. 7. Urbanization — Canada. 8. Conservation of natural resources — Canada. 9. Regional planning — Canada. 10. Agriculture — Canada — Planning. I. Pierce, John T. II. Title. III. Series.
HD205.F87 1982 333.76'0973 82-18424
ISBN 0-89291-149-2

Publication Supported by the A.A.G.

Graphic Design by D. Sue Jones and CGK

*Printed by Commercial Printing Inc.
State College, Pennsylvania*

Cover Photograph: U.S. Department of Agriculture, Soil Conservation Service, SCS-INF-164 Rev. 9-75, Modified by Kelly Culpepper and OJF.

Foreword

Fundamental to human life are agricultural systems that sustain us physiologically. Equally important to North American economy and society is the productivity of our agricultural sector. For many decades we have been aware of the humanitarian and political importance of North American food surpluses in a hungry world. We have also seen increasing stress in the farming community as rising costs of agricultural inputs have not been matched by market prices. During the 1970s energy costs soared, with farm impacts both in fuel prices and in the costs of energy-based inputs such as fertilizers, chemicals, and the embodied energy of machinery. The 1970s were the beginning of an era when cheap energy for agricultural intensification could no longer be anticipated as an easy solution to farmland losses.

Farmland preservation is a controversial issue. Some observers see losses of agricultural resources to urbanization, highway construction, and other non-agricultural uses as reaching crisis proportions; others point to apparently vast quantities of reserve or abandoned land, suggesting agricultural land preservation as unwarranted and costly. Assessment of the farmland preservation issue requires data on land conversion, evaluation of the quality of land resources for agriculture, projection of land needs for agricultural and non-farm uses, and prognostication of future gains in yields and magnitudes of national and international food demand. Only the future can tell us whether today's analyses were correct or misleading. But protagonists and antagonists of agricultural preservation agree that the issue is important, and that answers must be sought in the realms of science, politics, law, and practical experience.

In *Agricultural Land in an Urban Society* Owen Furuseth and John Pierce provide an international perspective on agricultural land preservation. Drawing on both Canadian and United States experience, they suggest some of the similarities and differences in farmland conversion processes, rates of agricultural land loss, policy alternatives, and preservation experience in North America. As geographers, Furuseth and Pierce bring important dimensions to the farmland preservation controversy — questions of not just *how much* land, but *where* and of *what quality?* Citizens of the United States will learn much by reflecting on the more pressing Canadian dilemma of limited prime agricultural land, much of it located in areas of greatest urban pressure. Canadians will find that similar problems exist in the northeastern U.S., and both audiences will benefit from understanding the potentials and limitations of policy strategies within the context of each other's legal system. We also trust that readers from abroad will benefit from this discussion, and add to it insights from their own experience.

Resource Publications in Geography are monographs sponsored by the Association of American Geographers, a professional organization whose purpose is to advance studies in geography and to encourage the application of geographic research in education, government, and business. The series brings contemporary research in the various fields of geography to the attention of students and senior geographers, as well as to researchers in related fields. The ideas presented, of course, are those of the authors and do not imply AAG endorsement.

C. Gregory Knight, *The Pennsylvania State University*
Editor, Resource Publications in Geography

Preface and Acknowledgements

North Americans live in an urban society. Despite evidence of a recent return to the countryside and decentralization of the population, the economic and political power of Canada and the United States still remains firmly in the cities. Most of those living in rural areas, including farmers, increasingly share the same expectations, values, and activities of their urban counterparts. If it were not for the genuine difference in the contribution of land to urban and rural 'production systems,' there would be little point in making a distinction between urban and rural. Viable agricultural production is dependent upon land's food-producing ability or its resource qualities, whereas no such condition applies to economic activities within urban areas. However, the important resource function of land is only slowly being accepted in North America. Basic private and public decision making structures, as evolved during the 20th century, remain weighted in favor of urban interests.

Our look at 'Agricultural Land in an Urban Society' focuses on the important issues surrounding the urbanization of agricultural land, the assessment of the relative effectiveness of policy reponses, and an assessment of opportunities for change in approaches toward farmland preservation. These individual foci of inquiry are, by necessity, painted with a broad brush. Details are far from absent, but our primary concern is to provide the reader with an overview of conceptual and empirical underpinnings central to a critical understanding of the societal significance and implications of farmland loss. Our approach involves considerable choice of information and interpretation of findings, both of which reflect the beliefs and assumptions of the authors. It is our contention that agricultural land resources are enormously important to the social well-being of North Americans. We argue that the long-term viability of the resource is in serious jeopardy unless we devise specific policies to improve soil management, to enhance farm incomes, and to redress imbalances between agricultural and urban competition for land. These views are also expressed in two recently published reports — the *National Agricultural Land Study,* and the *Global 2000 Report.*

The conceptualization and execution of this study were shared equally between us. JTP was primary author for Chapters 1, 2 and 3, while OJF was primary author for Chapters 3, 5 and 6. Chapter 7 was written cooperatively.

Needless to say, the study would not have been possible without the critical comments, support, and assistance of a variety of people. We would like to thank Greg Knight for his encouragement and the original reviewers of our proposal for their perceptive remarks. At the University of North Carolina at Charlotte special thanks to Suzanna Schwartz who provided valuable research assistance; to the UNCC Cartographic Laboratory which produced the bulk of the graphics for this volume; to Anita Mullen for Figure 12; and to Judy Patrum, Donna Rollins, Joyce Ingalls, and Denise Askland for their typing skills. At Simon Fraser, we are indebted to Moyna Bradley, Gwen Fernandes, Rosemarie Bakker, Nancy Burnham, and Ray Squirrell for manuscript preparation.

Finally, a very special thanks to Harriet and Jan for their support during the various stages of this project, and most importantly for their critical and practical advice and assistance.

Owen J. Furuseth
John T. Pierce

Contents

List of Figures

List of Tables

1

Urban Growth and the Competition for Agricultural Land

Until the 19th century, the growth of villages and cities was intricately tied to growth in local food production. Although trade and commerce weakened this relationship, the immediate hinterlands of human settlements provided the main life support system. Reliance upon animate energy further reinforced this dependency. Malthus, writing at the end of the 18th century, argued that since land area was fixed, increases in total population beyond a certain point would exceed the carrying capacity of the land and, thereby, create a situation of declining per capita food consumption. He also saw war, famine, and disease acting as 'positive checks' against runaway population growth. Few would have guessed at that time, or even almost half a century later, when Mill was pursuing similar questions, that increasing agricultural surplus, derived from land- and labor-augmenting technology, and the industrialization of society would significantly alter the calculus of population growth.

During the second half of the 19th century a dramatic increase in the concentration and growth of population were clearly emerging trends. Improvements in nutrition and sanitation extended average life expectancy, while a "rise in the technological enhancement of human productivity . . ." through the use of inanimate energy and specialization of production encouraged off-farm migration and urbanization (Davis 1965:13). If we accept Davis's (1965: 4-5) definition of urbanization as the proportion of the total population that is urban, then England and Wales were 50 percent urbanized by 1850 and approximately 75 percent by 1900. Similar proportions were not achieved in the United States until 1910 and 1960, respectively.

Urbanization in North America

Underlying the shift from a rural-agrarian society to an urban-industrial society in North America was increasing farm capitalization and productivity, off-farm migration, and changes in the economic climate of farming. Whereas 35 percent of the U.S. and 54 percent of the Canadian population were required for food production in 1910, only 4 percent were required by 1976. As incomes expanded beyond a basic minimum, proportionately less of an individual's budget was spent on food and proportionately more on manufactured goods. For example, in 1971 approximately 15 percent of disposable income of the average American consumer was spent on food compared with 65 percent of a consumer's income in developing countries (Heady 1976:77). Not surprisingly, the low income elasticity of demand for agricultural commodities in industrial societies (often referred to as Engel's Law) and the downward pressure on the price of agricultural commodities from surplus production have weakened the relative importance of the agricultural sector to the economy as a whole.

The Industrial Revolution produced an organizational revolution on an unprecedented scale. Increased productivity and efficiency were achieved by specialization of economic functions and labor. Specialization required increased labor cooperation and necessitated greater concentration of economic activities (Blumenfeld 1965:42). Agglomeration and urbanization economies became increasingly important locational factors for manufacturing and specialized service functions. Facilitating this cooperation and centripetal development were dramatic improvements in the integration of transportation systems. As a result, by 1920 relatively mature, well-integrated urban systems had emerged in both the United States and Canada. These systems, in turn, served as the loci for much of the subsequent growth in urban centers, growth which was generally at the expense of arable land. Ironically, these centers owed their existence and early growth to the surrounding agricultural base which they later consumed.

The most characteristic feature of urbanization in North America during the next fifty years was the increasing scale of urban development. Cities were transformed into metropoli as their economic base expanded and diversified. Growth became self-sustaining, the functional city engulfing the political city. The northeastern seaboard of the United States is the classic example of this phenomenon. The division between urban and rural diminished as formerly identifiable centers coalesced into a "unique cluster of metropolitan areas," which Gottman (1961:4) termed 'Megalopolis.'

Suburbanization

Urbanization of North American cities was followed by suburbanization. Growth in real income, democratizing effects of the automobile and telephone, and improvements in transportation extended mobility and facilitated urban expansion. In the U.S., 55.7 percent of the 1920-1930 population growth within Statistical Metropolitan Statistical Areas (SMSAs) occurred in central cities and 44.3 percent in suburban areas. In contrast, 22.5 percent of the 1960-1970 population growth of SMSAs occurred within central cities and 77.5 percent within suburbia (Berry and Kasarda 1977:171).

Economic and technological factors, although vital for suburbanization, must be placed within a wider context of social and economic forces stimulating the outward expansion of cities. Among these are federal tax and financial incentives, municipal land use controls, changes in the economic geography of manufacturing, and the importance of single-family home ownership (Holcomb and Beauregard 1981).

The post-World War II growth in North American suburbs and resulting consumption of rural and agricultural land was very much the product of federal housing and income tax policy. In Canada the National Housing Act provided insurance for lenders and the Central Mortgage and Housing Corporation supplied mortgage funds to selected income groups. In the United States the Federal Housing Administration and the Veterans Administration provided similar programs. Moreover, provisions in the U.S. federal tax code favored home ownership through mortgage interest deductions. This provision clearly subsidizes home ownership, particularly among middle and upper income groups, since tax savings are proportional to marginal tax brackets (Clawson 1971). Suburban municipal governments were also instrumental in encouraging urban sprawl through large-lot zoning, extension of public facilities and services, and low property taxes. No similar mortgage deduction scheme exists in Canada, and municipal actions are more closely controlled by provincial governments. These two factors account for some major differences between the U.S. and Canada in the rates and character of suburbanization and the consumption of agricultural land.

Counter-Urbanization

Beginning in the 1960s and clearly evident by the 1970s was a decline in the North American rate of population growth, including a pause in urbanization and a reversal in the concentration of people in relatively few metropolitan centers. Described as counter-urbanization or the rural renaissance, many non-metropolitan areas grew at nearly double the rate of their metropolitan counterparts (Berry 1980). The lower rate of U.S. metropolitan growth was due to out-migration from central city areas and a decline in suburban growth, particularly in the northern industrial areas. Since 1970, non-metropolitan areas in the U.S. have experienced net migration gains, a reversal of a 30 year trend. Although Canadian cities do not share this trend toward inner city decline, non-metropolitan and mid-size Census Metropolitan Areas (CMAs) have grown at a faster rate than the larger CMAs, particularly in central Canada. Decentralization and migration from larger to smaller centers again form the basis for this shift in population growth (Preston and Russwurm 1977: 18-24).

Despite lower population growth rates, the decentralization of population and higher growth rates of smaller centers ensure a continued spread of urban settlements and the consumption of agricultural land at levels comparable to times prior to counter-urbanization.

What are some of the factors responsible for decentralization? Berry (1980:43) argues that in the United States the migration of people from large metropolitan areas reflects new growth opportunities in rural areas — many related to resource and retirement development — as well as improved transportation facilities. In this connection he suggests that "the time-eliminating properties of long distance communication and the space-spanning capacities of new communication technologies are combining to concoct a solvent which has dissolved the agglomeration advantages of the industrial metropolis, creating what some refer to as an urban civilization without cities" (Berry 1980:52). Similar factors are at work reshaping the configuration of the Canadian urban system. Preston and Russwurm (1977:4) have noted a maturing of provincial urban systems accompanied by increasing importance of the service sector and resource development. Bourne (1978:134) has suggested the "push of labor surplus and the pull of labor demand," political uncertainty in Quebec, inflation, and the vicissitudes of commodity markets as factors affecting expansive urban growth in Canada.

Whatever the reasons behind the counter-urbanization process, the spread and growth of cities will continue. Given the historical bias of urban development toward the use of arable land and the superior competitive position of urban development against less intensive agricultural use (Smit and Conklin 1981:48), the consumption of agricultural land will continue to serve the locational and spatial requirements of urban areas. What, then, are the issues surrounding the urbanization of agricultural land?

Issues in the Urbanization of Agricultural Land

The potential and actual production of the North American 'food machine' is second to none. The enormous gains in post-war productivity are due in large measure to the industrialization of agricultural production, a process in which land- and labor-augmenting technology have combined to ensure successively larger surpluses (Heady 1976). The increasing price of agricultural land relative to other production inputs has encouraged a greater use of energy and a decline in land's relative importance in production. This substitution process treats land like any other commodity or capital good. Once freed from one economic activity — agriculture — land becomes available for another, urban development.

Many have argued that land cannot and should not be viewed within the narrow confines of factor inputs and production functions. Rather, land is very much a natural resource, offering many additional benefits to society. Moreover, the supply of land, unlike capital, is fixed. Thus, "the price paid for land does not finance the creation of new land" (Harriss 1980:128).

Benefits from Agricultural Land Resources

During the last twenty years, academics and planners have begun to appreciate the diverse functions inherent in agricultural land resources and the importance of sustaining them. Bosselman and Callies (1971:43-45) argued hat the notion of land as both a resource and a commodity is a highly novel concept:

> *Our existing systems of land use regulation were created by dealers in real estate interested in maximizing the value of land as a commodity . . . . A realization is growing that important social and environmental goals require more specific control on the use that may be made of scarce land resources.*

In discussing the benefits to society derived from actions to preserve agricultural land, Gardner (1977:1028-29) outlined four interrelated products:

> *"a) sufficient food and fiber to meet the nutritional requirements of a growing national and world population; b) local economic benefits that derive from a viable agricultural industry; c) open space and other environmental amenities that accrue chiefly to urban residents; and d) more efficient, orderly and fiscally sound urban development."*

Bryant and Russwurm (1979) explored similar themes, but within the context of the values inherent in agricultural resources. Farmland once incorporated into the agricultural production system produces economic goods and services. These goods and services are important elements in the structure and growth of the local, regional, and national economies. Less tangible but also important is "a long term 'value' inherent in the potential of the agricultural resource which affects the possibilities of alternative resource allocation at some future time" (Bryant and Russwurm 1979:123). It is the value of such potential, they suggest, which is most critical to farmland preservationists. Finally, Bryant and Russwurm stress that the open space function of agriculture, incidental to the production of food, is nevertheless highly valued in many European societies. In a study of farmers' response to urbanization in the U.S., three of the four goals underlying farmland preservation efforts were amenity or aesthetically related and only one was economic (Berry *et al.* 1976). The U.S. Environmental Protection Agency's (1978a) background paper on agriculture also stressed that aside from food and fiber production agricultural lands play an important environmental role. The EPA notes that farmland contributes to watershed protection, reduces pollution by trapping such compounds as ozone and sulphur dioxide, facilitates the disposal of sludge, acts as a buffer between expanding cities, and provides other amenity-related features (U.S. Environmental Protection Agency 1978a:5-8).

Market Imperfections and the Issue of Scarcity

Economists generally concede that agricultural land contains both a commodity and a resource function. However, few support either agricultural land preservation in the face of urban expansion or recognize market imperfections. The market is generally assumed to be competitive and, hence, will "allocate the socially optimal quantity of land to various competing uses" (Gardner 1977:1029). To support this belief, economists and non-economists, alike, point to the existence of a growing surplus of food on a quantitatively declining land base. Others argue that the existence of imperfections in the market is not, in itself, a *prima facie* case for intervention. There are often other equally efficient measures; opportunity costs associated with intervention often exceed the benefits; and government allocation is often just as imperfect as the market itself (Frankena and Scheffman 1980:11). To clarify the validity of these views and positions requires a systematic look at sources of market imperfection and the scarcity or adequacy of the land base for agricultural production.

Intertemporal Equity. Imperfections in the market often occur in an intertemporal or intergenerational context. Generally speaking, investment decisions and, in turn, the market allocation of land resources are heavily weighted in favor of the present. The private discount rate effectively determines the intertemporal allocation of resources. In its simplest form, the discount rate is the opportunity cost of capital. If the private discount rate is higher than the social discount rate then "the private market economy undervalues the well-being of future generations relative to society's criterion of fairness" (Frankena and Scheffman 1980:12). Consequently, agricultural land may be converted at an excessive rate, creating conditions of scarcity and high food prices for future generations.

The resulting trade-off between housing and agriculture is illustrated by McInerney (1981:51-54) in a simplified land use model (Figure 1). The intersection of the marginal net benefit schedules (MNB) for housing and agriculture represents a presently optimal land use policy. Clearly allocation of Q_H units of agricultural land to urban uses is technically irreversible, and future agricultural services cannot be reestablished. However, land currently providing agricultural services can in the future be used for either housing or agriculture. Hence, "care must be taken in the current period . . . to ensure that no transformation of land is made which, although optimal for the current period, results in insufficient land being available for the provision of agricultural services in the future" (McInerney 1981:52). If, in fact, more than $S-Q_H$ units of land are required for agricultural production in the future, and the incremental net benefits from land converted to housing are less than the increment of lost future benefits from foregone agricultural production, then future agricultural benefits have been excessively discounted and the allocation of housing space has not been optimal.

Frankena and Scheffman (1980:13) argue that while government intervention into the land market may be necessary to protect the interests of future generations, limited information on social discount rates and future values of land in different activities would make intervention very difficult.

Resource Scarcity. Closely related to intertemporal equity is the issue of scarcity of natural resources in relation to sustained economic growth. Stiglitz (1979:40) has suggested conditions that must occur together before there is a "meaningful natural resource problem:" a resource must be in limited supply relative to current use; it must be non-renewable and non-recyclable; it must be essential with no existing or potential substitutes; and efficiency improvement beyond a certain point must be impossible. To

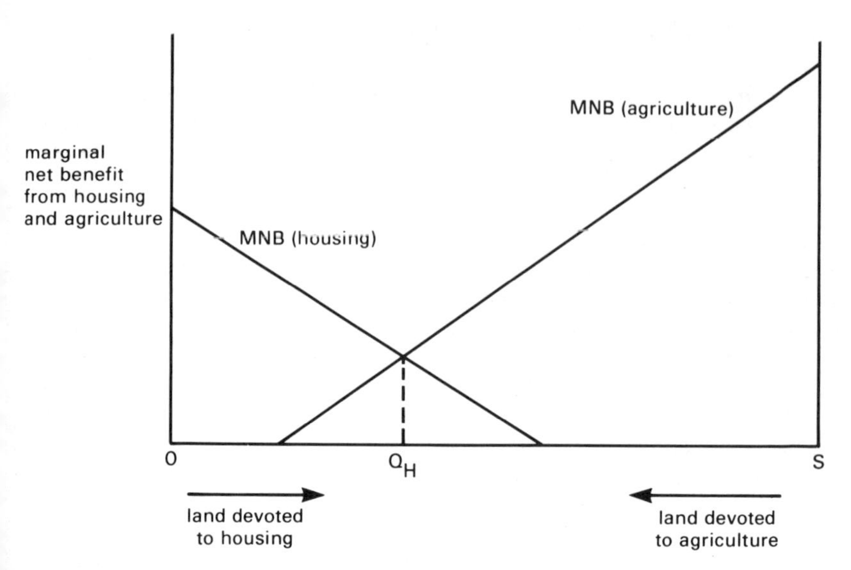

FIGURE 1 OPTIMAL ALLOCATION OF LAND RESOURCES (Modified by the authors from McInerney 1981). O, zero supply of land; S, maximum supply of land; Q_H, quantity of land for housing.

some observers these distinctions are a form of sophistry equivalent to Schumacher's (1979) warning that we are merely re-arranging the deck chairs on the Titanic. To others they provide a standard against which to examine the adequacy of the resource.

While agricultural land resources are not in limited supply in relation to current use, they are largely "inexhaustible but nonaugmentable" (Stiglitz 1979:38). However, issues of equity, efficiency, and substitutes are less straightforward. It is clear that land is essential for agricultural production since it combines both space and productive capabilities. How far can technology extend the substitution of nonland inputs for land inputs? Is there, in other words, a limit to the "technological fix?" To date, the elasticity of substitution has been high and constant, owing to improvements in the efficiency of farm production. But can we assume that this elasticity will remain constant? There is reason to believe that continued population growth accompanied by the territorial expansion of cities and the inability of technological change to offset this loss of land will contribute to scarcity and, hence, to a decline in the elasticity of substitution. Already, the rate of production increase of a number of cash crops has either levelled or decreased over the last ten years (Crosson 1977). With the exigencies of water and energy availability and soil erosion, there is little chance that the conditions which gave rise to food surpluses in the past will continue indefinitely in the future.

Government Intervention

There is no definitive answer to the question of whether the North American land base will be adequate for food production. From the standpoint of present sufficiency, little evidence suggests that a true resource problem exists and government intervention is warranted. Nevertheless, the future remains unclear, since we are dealing with a finite resource whose productive potential is dependent on finite energy resources and whose other services to society are likely to become even more important. Perhaps if government intervention is warranted, it is in response to future resource potential as opposed to problems of existing food production (Bryant and Russwurm 1979: 134).

Ciriacy-Wantrup (1965) has argued for the adoption of some safe minimum standard of conservation in the use of certain flow resources to avoid "immoderately" large losses and a decrease in society's development options. "The economic rationale of a safe minimum standard is based upon the proposition that the costs of maintaining it are small in relation to the possible losses which irreversibility of depletion might entail" (Ciriacy-Wantrup 1965:581). The definition of the minimum standard for high capability agricultural land would depend upon identifying a critical zone for food production which would be required under future demand and technological conditions. The critical zone would be much more modest than a theoretical social optimum. Establishment of a safe minimum standard would act as an objective standard or goal against which state, provincial, and local governments could either evaluate the success of their farmland preservation programs or determine the need for such programs. With a safe minimum standard in place it would be possible to "explore whether and how the social optimum in the state of conservation could be more closely approximated in [a] step-by-step fashion" (Ciriacy-Wantrup 1965:581).

The alternative to public involvement is reliance on the marketplace to allocate land uses. The past performance of the land market has not been reassuring when one considers the scale and impact of its imperfections. These imperfections can be traced to uncertainty, lack of information, externalities, and the presence of collective goods. Lack of information, for example, seriously constrains efficient allocation of resources. Contributing to this condition is what Stiglitz (1979:51) calls the absence of futures markets for land or, in other words, the problem of forecasting future demand and supply conditions. The difficulty in obtaining data, the diversity in assumptions regarding technological change, and the cost of research are all impediments to accurately assessing future resource needs.

Externalities are a serious problem in areas undergoing urbanization. Rural nonfarm residential development creates problems of congestion, pollution, farm fragmentation, weed proliferation, and decline in the service infrastructure for farming. The farmer is not the only person to bear the cost of inefficient development. Servicing fragmented and/or sparsely populated residential development is costly both financially and aesthetically (Real Estate Research Corporation of Chicago 1974). Externalities have traditionally been subject to local control in the form of subdivision regulations, zoning bylaws, and local ordinances. In most cases effectiveness has been limited.

Although support for government intervention into the allocation of land resources because of uncertainty, lack of information, and externalities has been relatively weak, the reverse has been the case for collective or public goods. Most economists concede that the market is generally unable to supply public goods because the supply price is zero. Since the absence of private ownership precludes the collection of rents or the recovery of costs there is no incentive to produce public goods. Consequently, one of the arguments for the protection of agricultural land is grounded in the notion that the land has qualities of a public good, although impure. Agricultural land provides open space and amenity values much like forestry. Stiglitz (1979:38) refers to this type of good as a 'publicly provided private good'. Similarly, agricultural land could be considered privately owned but publicly enjoyed. Whether a growing urban society can sustain its per capita consumption of this good depends very much upon the substitutability of the resource in a spatial sense. Complicating this question is the enormous variation in the amenity quality of the open space. Given the importance of the amenity function of agricultural land and the inability of the market to ensure what society might consider adequate supply, some sort of public protective action is likely warranted. Action is particularly needed in areas where the opportunity for substitution is limited and quality of the resource is unique.

2

Dynamics of Land Conversion

Traditionally the impact of urbanization on agricultural land use includes direct and indirect effects (Coughlin *et al.* 1977). Direct effects of urbanization usually refer to the conversion of farmland to urban uses, whereas indirect effects refer to the less tangible, but nevertheless significant, changes in farmer attitudes, the structure of farming, land values, and rural infrastructure. After examining several models of the rural-to-urban land conversion process and the sequence of events which normally preceed conversion, we will consider some legal and constitutional questions surrounding public intervention into the land market.

Understanding Land Conversion Processes

The search for regularities in the rural-to-urban land conversion process can proceed in several directions, both theoretical and empirical. In addition, studies may focus on changes in specific cities through time or on comparison of cities across the urban hierarchy.

Population Density in Urban Areas

The density-distance decay model for urban areas has long been accepted by geographers and other urban analysts. The model suggests that population density falls with increasing distance from the city center. The exact form of this gradient has been described in a number of ways. Newling (1969) argued that whereas population density declines exponentially with distance from the city center, the rate of density growth is a positive exponential function of distance from the center. Therefore, the rate of density increase is inversely related to density. The actual sequence of urban density increase has been empirically examined by Blumenfeld for Philadelphia (1959) and Toronto (1967). Population densities within a series of concentric zones were examined over a 50 year period. The research findings showed that population density moved outward at a mathematically predictable function of the percentage growth of the total area population during the given period and the mileage of the density zone from the center (Blumenfeld 1967:323-324). Blumenfeld characterized this growth process as a "tidal wave of metropolitan expansion."

In a more recent longitudinal study of Toronto, Latham and Yeates (1970) found that density gradients had assumed a quadratic negative exponential function of distance from the city center. Their analysis revealed an outward movement of the density profile. The authors also characterized the change in urban density gradients,

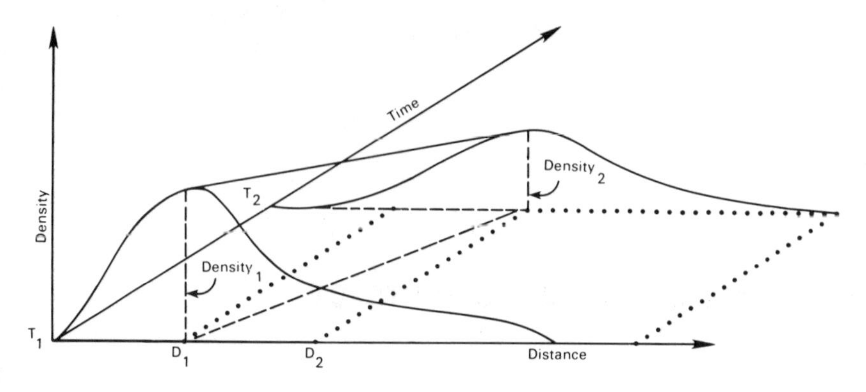

FIGURE 2 WAVE ANALOGUE OF URBAN POPULATION DENSITY (Boyce 1966; reproduced by permission of B. C. Geographical Series).

the most important features of which were declining central city densities; densities increasing over time with distance from the city center; and, with this outward displacement of the density profile, an expansion in the radius of the urbanized area.

Boyce (1966) has conceptualized this change of density over both space and time with a simplified version of Blumenfeld's wave model for urban growth (Figure 2). This wave analogue, along with perspectives of Latham and Yeates, provides a useful framework for understanding some of the trends in urbanization. Decentralization and the growth of population are very important components whose rates of change affect the radius of the city and, as a consequence, per capita land consumption.

Catastrophe Theory

Models of spatial expansion of the city often depict population growth and distribution as continuous. This continuity may be more a function of the scale of measurement than any inherent regularity. Urban growth is a cyclical and discontinuous phenomena dependent upon business cycles and, to a lesser extent, planning processes. To gain some understanding of the discontinuous nature of urban growth and the consumption of agricultural land we enlist the support of catastrophe theory.

Developed by Thom (1975), catastrophe theory is a method of describing, explaining, and predicting discontinuous phenomena, principally those possessing systems properties. Thom presented a number of archetypal forms (seven in number) which he called elementary catastrophies (Zeeman 1976). In modelling the physical expansion of cities, the geometric form most useful is the "cusp" catastrophe surface.

The model in Figure 3 assumes that population change at the urban fringe is controlled by two factors: planning regulations and demand for urban space. Various combinations of planning intervention and land resource demands, both indicated on the control surface, produce a range of density outcomes on the behavior surface. These outcomes take two contrasting forms — continuous and discontinuous. In the continuous case, sprawling and haphazard development reflects the relative absence of planning. An increasing demand for space gradually and incrementally shifts an area on the urban fringe from a predominantly rural agricultural (low density) to urban residential (high density) environment. In contrast, nucleated and organized development reflects systematic planning approaches which tend to concentrate development and to express demand suddenly. A rural agricultural area is catapulted to urban residential status. The population within the area would then level off representing a stable equilibria until the demand for space is given expression through planning approval or the extension of services.

Each of these forms of development contain different implications for the viability of agriculture, cost of urban expansion, and the choice of urban residents. Concentrated, planned developments reduce the alienation of farmland, minimize uncertainty and negative externalities for farmers living in the vicinity of urban areas, and reduce servicing costs for municipal governments. Against these advantages, the new resident is likely to be faced with higher per-unit land costs and greater restriction in locational choice than otherwise occur with sprawling unplanned development.

Conversion-Size Rule

The models discussed provide a framework for understanding some of the dynamics of urban expansion and the conversion of rural land at the metropolitan scale. If there are regularities at this scale, are there also regularities among cities in terms of the consumption of rural land? Do land conversion rates throughout the urban hierarchy respond systematically to population size and population change?

That the physical structure of cities varies systematically with population size is a well-understood feature of urban systems. Best and collaborators (1974:206) described this relationships as a *density-size rule:* "as the population size of a settlement increases, land provision declines exponentially (i.e. the density of development rises at a decreasing rate)." The rule can also be expressed in terms of the relationship

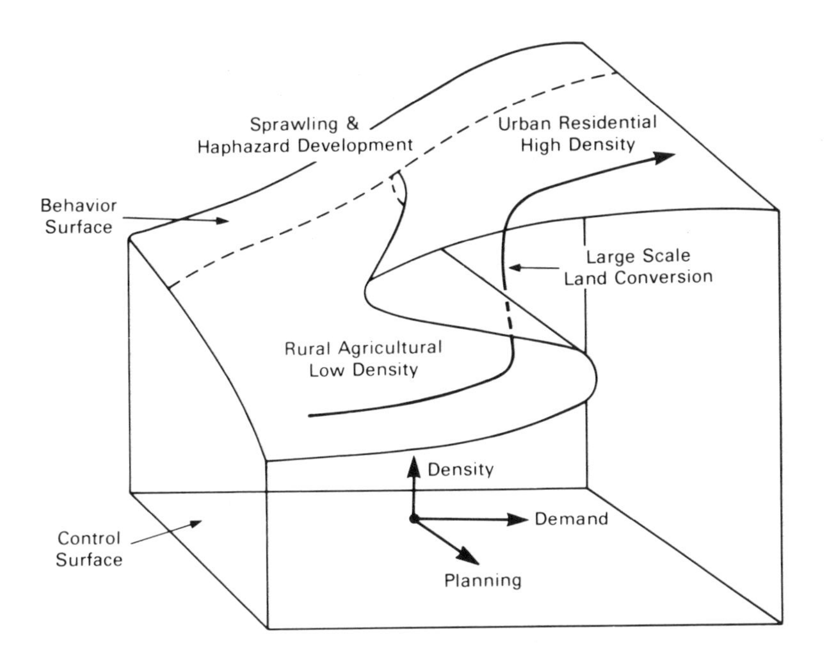

FIGURE 3 A CUSP CATASTROPHE MODEL OF RURAL-TO-URBAN LAND CONVERSION.

between population size and two other variables: developed area and acres per person. Most studies of the relationship between density and population size have been more concerned with "the end or average product of the growth process than with the mechanics or marginal product of the growth process" (Pierce 1979:334).

If the net change in urban land use throughout the urban hierarchy (reflecting the quantity of rural or agricultural land consumed during territorial expansion) is related to change in population, then we can identify a relationship similar to that which exists between developed area and population size. We can call this the *intercity conversion size rule* in which: "for any given time interval, as the growth in the city's population increases, developed area will also increase but at a decreasing rate" (Pierce 1979:337).

The decline in the per capita consumption of land with increasing population change and, by extension, population size nevertheless indicates an increase in the demand for land in relation to its supply. Clearly, there are exceptions to this rule which emphasize the influence and importance of other variables, not the least of which are city age and structure (Winsborough 1962). Nevertheless, as cities increase in size, the rate of increase in density diminishes and, as population growth increases, the rate of decrease in per capita consumption of land diminishes. As a result of these trends, proportionately less agricultural land is lost as population growth increases. Higher-density urban areas absorb more population growth per unit of expansion in area than lower density communities.

Indirect Effects of Urbanization

Negative Effects

Much of the furor over the urbanization of agricultural land is due not so much to land conversion to urban use as to far more extensive indirect effects. The ultimate source of most indirect impacts is population growth in rural and semi-rural areas (Coughlin *et al.* 1977). Indirect effects may begin with a change in attitude and behavior regarding the future of farming. Growth in part-time farming, idling of agricultural land, reduction in capital investment, and change in the mode and type of production often signal the anticipation of urban development. The physical process of urban expansion usually comes much later.

Sinclair (1967) was the first to conceptualize the relationship between urban anticipation and the value of agricultural investment. He argued that as urban anticipation decreased with distance from the city, agricultural investment would increase, as would the value of agricultural production. Thus, the price of land would reflect its agricultural rather than urban potential. Sinclair, however, ignored the role of transport costs and the varying investment horizons required for production by different activities.

In an extension of Sinclair's model, Bryant (1973) attempted to incorporate the influence of two cost components: the variation in transport costs and the costs of urban expansion which he termed "capital fixity" (Figure 4). The shape of each activity's bid-rent curve reflects assumptions regarding the degree of capital fixity, amortization, importance of labor, transport costs, and the time left for agriculture (TLFA). The extension of each of these curves (broken lines) represents the economic rent that each activity could earn in the absence of urbanization pressures. Since the time left for agriculture increases with distance from the city (or inversely with the urban anticipation), bid-rent curves for activities with high capital fixity will also increase with distance

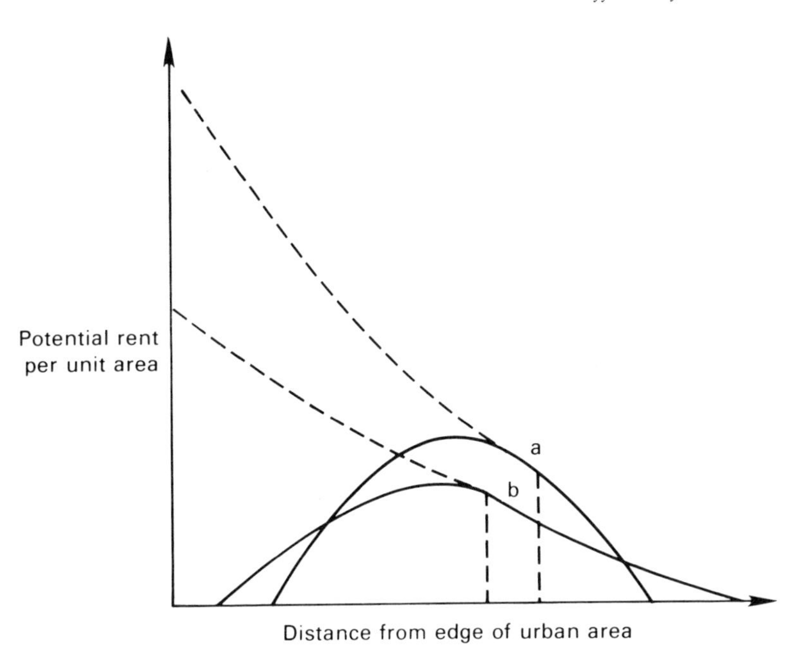

FIGURE 4 POTENTIAL RENT CURVES FOR AGRICULTURAL USES WITH TRANSPORT COSTS AND A LINEAR TLFA SCHEDULE (Courtesy of C. R. Bryant; Bryant 1973). TFLA, time left for agriculture.

(curve a). The peak of each curve occurs when time available for agriculture approximates normal conditions and the TLFA component becomes a constant. The downturn in the value curves reflects the increasing cost of transportation. Despite the fact that the model does not incorporate the role of rental land, or the fact that urban anticipation may not vary systematically with distance, Bryant's extension offers a useful scheme for understanding the persistence of some activities undergoing strong urbanization pressures as well as the growth of high investment activities such as dairying beyond areas of strong urbanization pressures (Berry 1978).

Underlying the change in attitudes and structure of farming brought about by anticipated urbanization is a complex set of interacting factors which are closely tied to population growth and redistribution. Numerous studies have identified these factors as indirect effects. However, they can as easily be viewed as 'intervening' forces of change which act in a non-linear interactive manner, displaying symmetrical and asymmetrical relationships much in the same way as in Myrdal's (1957) model of circular or cumulative causation. These intervening forces can be subdivided into three categories: developmental, status and spillover, and resource hinterland.

Developmental Forces. Among the conditions associated with the physical conversion of land from rural to urban use are farmland fragmentation, decline in the agricultural service sector, dramatic rise in land values and speculation, increase in private land banking, and idling of farmland. Not only are these the most visible products of development, but also the most influential forces affecting the behavior of the farmer since they pose the greatest threat to the continuity of the agricultural environment. Fragmentation of land is due to weak land use planning and the uncontrolled provision of services, whereas the decline in the service sector is directly related to the physical loss of farmland and the unwillingness of farmers to invest in their enterprises. Although land speculation occurs throughout the "urban shadow," its frequency and impact is greatest within the vicinity of existing development, as is the creation of private land banks and the decline of actively farmed land.

Status and Spillover Forces. Coughlin and colleagues (1981:53) have made the observation that farmers and other rural residents "begin to lose their political, economic, and social status in the community in which they live as they become out-numbered by people from different economic, political, and social interests." As the political complexion of an area changes, the ability of farmers to preserve the rural *status quo* diminishes. Increasing urbanization translates into environmental externalities. Farmers may be threatened by air pollution; the water table can be lowered through the proliferation of large, unserviced lots whose only source of water comes from wells; crops and equipment may be damaged by vandalism. County or municipal authorities are often called upon to regulate the 'normal' activities of farmers. These regulations take a variety of forms but restrictions in road access for slow-moving farm vehicles, noise and odor ordinances, and spraying controls are perhaps the most serious impediments to farming. Moreover, increased farmland value restricts farm expansion and, in the absence of special legislation, subjects the farmer to increased property, school, and infrastructure taxation. Spillover effects can be interpreted even more broadly to include the 'pull' of off-farm employment. Urbanization offers new opportunities for many farmers facing low income potential. The shift to part-time farming is an important component of the changing structure of agriculture.

Resource Hinterland Forces. Cities are served by their hinterlands in a variety of ways. Not only does the urban fringe provide excellent recreational and amenity values but also acts as a source of important physical resources for urban residents. Water reservoirs, sand and gravel operations, and utility and transportation corridors are vital to the growth and development of individual urban centers and the linkages which define urban systems. The demands by these activities for rural land resources are enormous and very often incompatible with viable farming. Unlike the previous two forces which result from population growth in the rural fringe, the resource hinterland is affected by growth in the entire urban area as well as a change in demand for services by the existing population.

Finally, it should be emphasized that while developmental forces are generally limited to the fringes of urban areas, status and spillover and resource hinterland are far more pervasive influences on land use decisions at locations distant from the scene of active urbanization.

Positive Effects

Typically, population growth in the urban fringe is likely to be incompatible with viable farm operations. There are instances, however, when population growth in agricultural areas has been beneficial to the well-being of farmers and to the general farm environment. For example, an enlarged population base often improves the quantity and quality of services and provides for greater opportunities in marketing products. Hobby farms provide a stabilizing element in an otherwise rapidly changing rural environment, preserving such amenity features such as wooden barns, historic houses, and tree-lined lanes. Bryant and Russwurm (1979) pointed out that the negative effect of urbanization upon farming operations is a function of changes in population density, the ratio of farm to non-farm residents, and the rapidity of change. Conflicts and incompatibilities become significant once a critical threshold is reached. The precise definition of this threshold is difficult, however. It depends on such intangible qualities as urban-rural cooperation, respect for the rural environment, and the strength of rural institutions.

The Land Market in the Urban Fringe

The suburban growth of North American cities has been likened to the process of colonization in that "both reflect an imbalance between resources, especially between land and population" (Mills 1973:54). Although the physical supply of land is fixed within any given urban region, its (economic) availability for development is closely related to distance or accessibility factors. The growth in transportation routes is at the heart of distance/accessibility relationships, changes in relative location, and, ultimately, the price of land. Other important determinants of price include physical features, amenity values, high quality educational facilities, lot size, and zoning controls. Clearly, the supply of land for development will change in response to changes in price. In theory, at least, land will become available when an owner feels that the market value equals or exceeds future value discounted to the present.

Despite the attractiveness of these mechanistic notions of growth in land supply for development, many political, institutional, and behavioral factors influence the availability of land. The pattern of leapfrog development around many U.S. cities belies many commonly held economic explanations. The exercise of monopoly rights over a single location and the divergence between use and exchange value of land further compound the problem. Equally significant is the importance of public service facilities and subdivision regulations. These public policy decisions can control the location, rate, and intensity (or form) of development, or, in short, shape the environment of choice. Of course, the existing profitability of farming, the farmer's perceived imminence of urbanization, and spillover effects are important ingredients in slowing or accelerating the sale of farmland to developers or speculators.

If one wishes to influence successfully the location and pattern of land conversion, then one must understand the principal decision makers and the linkages that exist among them. The complexity of decision making should not be underestimated by policy makers. The land market is characterized by numerous individuals and companies with diverse objectives, abilities, and financial power.

In their model of the conversion process, Kaiser and Weiss (1970), for example, suggested that land passes through a sequence of stages in preparation for development, with each stage a product of a complex set of decisions and decision agents. Although the shift from agricultural to urban is of greatest importance for our study, the full range of stages are significant since they are all linked, interdependent spatially and temporally. To explain the transition of land from one state to the next, Kaiser and Weiss introduced three types of decision factors: contextual, decision agent, and property characteristics. Contextual factors are general socio-economic conditions and public policy variables which include a wide range of municipal infrastructural services and land use regulations that affect the conversion of land. Property factors are those physical and locational characteristics associated with an individual parcel of land. Decision agent characteristics refer to the actors in the development process and their differing motivations and expectations. Having a significant impact on all three factors is the last component of the Kaiser-Weiss model — local public policies such as taxation rates, building codes, and annexation provisions. The influence of these policies on land transactions varies according to the nature and composition of the three previous decision factors.

Building upon the work of Drewett (1971) as well as Kaiser and Weiss (1970), Martin (1975) conceptualized the land conversion process according to a number of stages of development which disclose a progressive shift toward an urban landscape (Figure 5). Unlike earlier studies, Martin subdivides decision agents into primary and secondary, the former encompassing the individuals making the land transactions, the latter encompassing those individuals who comprise the service and regulatory structure through which transactions normally take place. Associated with each stage and set of decision making agents are the incremental changes in cost and price of land (price 1 . . . 5), the total of which (price 6) reflects the accumulated cost in the previous five states.

Intervention into the Land Market

Public intervention into the land market has taken many forms but, generally speaking, it can be considered to be the use of 'extra-market means' to determine land uses for the public good in both the present and future. Through outright expropriation, or the selective control of the 'bundle of rights' associated with the land, all levels of North American government have influenced the allocation of land. That institutional action often limits the actual or potential exchange value of land in a society which upholds the principle and rewards of private property ownership is the source of much of the opposition to government involvement in the land market. The critical issue is whether compensation, as a result of government intervention, is payable. This, in turn, is dependent upon the establishment of "taking."

In the United States, the Fifth and Fourteenth Amendments of the Constitution read: "No person shall be deprived of life, liberty or property, without due process of law; nor shall private property be taken for public use without just compensation" (Fifth Amendment). ". . . Nor shall any state deprive any person of . . . property, without due process of law" (Fourteenth Amendment).

While these amendments provide the general legal principles for the protection of property rights, they do not define how much regulation of land is permissible before it amounts to "confiscation" or "taking." Many early decisions on taking were rooted in the notion that an individual either had the right to use property for its highest and best use or, at minimum, was permitted development subject to planning regulations such as density controls and use restrictions. Under this commodity view of land, denial of a reasonable return through certain zoning regulations or development controls would most likely have been considered to be taking without compensation.

Bosselman (1976:68) argued that in recent years courts and legislatures have moved toward a different view in which "the owner of land has the right to make some economic use of the land but not necessarily the right to develop it" and, in some cases, "the owner of land does not have any rights to use it or develop it any particular fashion." Thus, the restriction on development recognizes the natural resource status of land (Sobetzer 1979). An analysis of the taking issue by Bosselman and Callies (1971) revealed that, generally, courts have not ruled taking if the land has certain environmental features which require it to remain in its present condition or when ". . . the

Stages of Development

	Non-Urban Use	Non-Urban Use: in Urban Fringe	Urban Interest	Active Purchase of Raw Land	Active Development	Active Purchase of Developed Land
Description of Stage	In agricultural or other non-urban use	Change in use or degree of intensity of use	Decision agent recognizes development potential	Decision agents negotiate land sales transactions	Physical Development of Land	Land purchase by property user
Decisions	No decision to sell in present or future	Consideration of future sale	Present use viewed as transitional	Purchase land	Develop land	Purchase home
Primary Decision Agents	Farmer	Farmer Land dealer	Farmer Land dealer Developer	Developer	Developer Builder	Builder Family
Secondary Decision Agents		Financier	Financier	Financier Lawyer Realtor Planner	Financier Lawyer Planner	Financier Lawyer Realtor
Financial Support	Unchanged	Farm credit	Negotiation with lender	Vendor mortgage	Development mortgage	Residential mortgage
Price of land	price 1					
Improvements		price 2				
Speculation			price 3	price 3		
Public Improvements					price 4	
Contruction						price 5
Appreciation						price 6
Total Development Costs						price 1 ... price 6

FIGURE 5 STAGES OF DEVELOPMENT (Courtesy of L. R. G. Martin, University of Waterloo and Environment Canada, after *Land Use Dynamics on the Toronto Urban Fringe;* Martin 1975).

restrictions are based on some overall comprehensive planning approach . . ." (Bosselman 1976:69). Hence, from either a natural-resource or public-interest perspective, it is possible to legislate limited or no development without the statute being invalidated as a taking without compensation.

Although the legal basis of both Canadian and U.S. law are found in English Common law, the taking issue in Canada has been viewed differently by the courts. No constitutional principle in Canada guarantees property rights or compensation for those rights when taken by regulation. Parliament is the supreme law maker and, according to Challies, 'the only guide to what Parliament may do is what Parliament has done' (quoted in Wood 1977:70). Wood (1977:70) argued that "since there is no constitutional right to compensation for a compulsory acquisition (expropriation) of land in Canada, then it must follow there is no constitutional right to compensation when what is being affected is not property, per se, but the right to use the property in a certain way."

In place of constitutional guarantees, provincial governments have passed expropriation acts to provide for the 'right' to compensation which, in some cases, extends to reductions in land values. But generally speaking, these rights are cleary defined such that there is no general right to compensation where land use regulations affect, for example, land values. This fundamental difference in legal structure and conventions leads not only to different attitudes but also to different actions to regulate urbanization and protect agricultural land between Canada and the U.S.

3

Land Resources for Food Production

In considering the value and place of agriculture within an urban society, one initial task is to identify those lands which are most critical, having the greatest potential agricultural value. This assignment is, of course, a prerequisite for any planning program or land use policy designed to foster agricultural development. Initially focusing on the identification of valuable agricultural land resources in Canada and the United States, we subsequently assess the geographic and economic distribution of these resources in North America.

Defining Valuable Farmland

There are a variety of methods for classifying land according to its value or importance for agriculture. Among the approaches presently used are identification of land currently being farmed, calculation of the economic or monetary yield attainable from a parcel, or development of an objective assessment of production potential using soil and climatic data (Geier 1980:671). Current farm status and economic yield are inadequate criteria when used alone, however. Farm status may result in the identification of marginally productive land, while economic yield may fail to identify potentially valuable land owing to problems with inaccurate data or fluctuations in the price of agricultural products (U.S. Soil Conservation Service 1975). Additionally, long range national concerns require that we know which land needs the least energy, fertilizers, and other external inputs and has the fewest environmental hazards (Dideriksen and Sampson 1976).

The most widespread and respected classification scheme is based on soil characteristics, that is, the inherent suitability of the land resource for agricultural purposes. The primary advantages of this approach are the relative ease and low cost of classifying a parcel and the permanence attached to the classification. The national governments of both Canada and the United States have systematically inventoried soil resources for several decades. Presently, there is almost complete coverage of both nations' agricultural regions. Using the soils data already collected, agricultural suitability is easily determined. Moreover, the type of classification based on soil characteristics is not subject to rapid obsolescence.

The foremost examples of classification based on production potential are the Canada Land Inventory's Soil Capability Classification for agriculture (SCC) and the U.S. Soil Conservation Service's Land Capability Classification (LCC). Each classification has undergone a rigorous review and testing program and is highly regarded. Both have been operational since the mid-1960s. The SCC is the standard classification system of soil capability across Canada, while the LCC enjoys the same position in the United States.

The principal criteria in both systems are adequate moisture; capacity to hold moisture without waterlogging; susceptibility to erosion; freedom from excessive alkalinity, acidity, and salinity; stoniness; and climatic variables (Geier 1980). Using these criteria the SCC group soils in seven classes, with 13 subclasses used to identify the specific limitations of the soil resource. Each class is identified by an Arabic numeral (Table 1). The LCC breaks down the soil resource into eight classes, using Roman numerals, with four subclasses (Table 2). In both systems the agricultural suitability of the land decreases as the class number increases.

Prime Farmland

Within the context of the land capability classification, the U.S. Soil Conservation Service has developed the concept of prime agricultural land, land which is or could be highly productive and suited to sustained, intensive cultivation. Criteria for prime lands

TABLE 1 CANADA LAND INVENTORY SOIL CAPABILITY CLASSIFICATION FOR AGRICULTURAL USE

CAPABILITY CLASSES

Class 1	Soils in this class have no significant limitations in use for crops.
Class 2	Soils in this class have moderate limitations that restrict the range of crops or require moderate conservation practices.
Class 3	Soils in this class have moderately severe limitations that restrict the range of crops or require special conservation practices.
Class 4	Soils in this class have severe limitations that restrict the range of crops or require special conservation practices or both.
Class 5	Soils in this class have very severe limitations that restrict their capability to produce perennial forage crops, and improvement practices are feasible.
Class 6	Soils in this class are capable only of producing perennial forage crops, and improvement practices are not feasible.
Class 7	Soils in this class have no capability for arable culture or permanent pasture.

CAPABILITY SUBCLASSES

Subclasses are divisions within classes that have the same kind of limitations for agricultural use.

C	Adverse climate
D	Undesirable soil structure and/or low permeability
E	Erosion
F	Low fertility
I	Inundation by streams or lakes
M	Moisture limitation
N	Salinity
P	Stoniness
R	Consolidated bedrock
S	Adverse soil characteristics
T	Topography
W	Excess water
X	Cumulative minor adverse characteristics

Source: Canada, Department of Regional Economic Expansion 1965.

are based on high quality soil characteristics. Viewed in terms of the LCC, prime agricultural land encompasses Classes I and II land and some Class III land (Diderik-sen and Sampson 1976:196).

The Canada Land Inventory's SCC does not formally distinguish prime agricultural land. However, the system was modified by Environment Canada (1972) to incorporate this concept. Prime agricultural land in Canada has been subsequently defined as soils in Classes 1, 2 and 3 (Simpson-Lewis *et al.* 1979). Generally, this designation has been accepted by provincial and local agencies throughout Canada.

Agroclimatic Resource Index

The use of soils data to classify the physical potential of land for agriculture is well developed, but an important variable still widely neglected is climate. Not surprisingly, Canada, with much harsher climate, has done far more to analyze and classify climatic parameters for agriculture. The Canada Land Inventory has noted that ". . . Climatic information is needed to assist in the understanding of land productivity as related to the development of Canada's resources" (Chapman and Brown 1966:1).

In an effort to devise a method for comparing the climatic resource base in different parts of Canada, Williams (1975) evolved the Agroclimatic Resource Index (ACRI). The basic parameter of the ACRI is the length of the frost free growing season, as modified

TABLE 2 U.S. SOIL CONSERVATION SERVICE LAND CAPABILITY CLASSIFICATION

CAPABILITY CLASSES	
Class I	Soils with few limitations that restrict their uses.
Class II	Soils which have moderate limitations that reduce the choice of plants or that require moderate conservation practices.
Class III	Soils that have severe limitations that reduce the choice of plants, require special conservation practices, or both.
Class IV	Soils having very severe limitations that reduce the choice of plants, require very careful management, or both.
Class V	Soils that are not likely to erode but have other limitations, impractical to remove, that limit their use largely to pasture, range, woodland, or wildlife.
Class VI	Soils having severe limitations that make them generally unsuited to cultivation and limit their use largely to pasture or range, woodland, or wildlife.
Class VII	Soils which have very severe limitations that make them unsuited to cultivation and that restrict their use largely to pasture or range, woodland, or wildlife.
Class VIII	Soils and landforms with limitations that preclude their use for commercial plants and restrict their use to recreation, wildlife, water supply, and esthetic purposes.

CAPABILITY SUBCLASSES	
a	Erosion
w	Water in or on the soil interferes with plant growth or cultivation
s	Shallow, droughty, or stoniness
c	Too cold or too dry

Source: U.S. Soil Conservation Service 1966.

FIGURE 6 CANADIAN AGROCLIMATIC RESOURCE AREAS (Map courtesy of G.D.V. Williams, Environment Canada. Map reproduced from *Canada's Special Resource Lands;* Simpson-Lewis 1979). See text for interpretation of values.

by moisture limitations and summer heat or degree-day data (Simpson-Lewis *et al.* 1979:15). The resulting index produces values ranging from less than 1.0 in the north, beyond the limits of normal agricultural production, to 3.0 in portions of southern Ontario (Figure 6). Lands with ACRI values of more than 2.0 have been labelled as critical agroclimatic resource areas. It has been suggested that special policies and practices are needed to wisely manage these irreplaceable resources (Simpson-Lewis 1979:18).

Through the combination of soil capability and climatic data, the Canada Land Inventory is able to identify and compare the agricultural potential of diverse land resources throughout Canada. There is no equivalent ystem in the United States.

Existing and Potential Agricultural Resources

Over the past thirty years, North America has emerged as the most important agricultural region in the world. Our abundant agricultural resources, particularly the prime agricultural land base, have contributed much to Canadian and American economic growth and eliminated the need to rely on outside sources for basic food and fiber requirements.

Canada's Agricultural Lands

The second largest country in the world, Canada has land surface area of approximately 2.3 billion acres (922 million ha), but only limited land capable of supporting economically viable agriculture. Eighty-six percent of the country has either no capability for agriculture or has not been classified using the Canadian Land Inventory (CLI). Encompassed in this category are areas which, because of extreme climate, lack of soil, ground water conditions, or previous commitment to more intensive uses, are unsuitable for agriculture. Of the remaining 13 percent, approximately 2 percent is marginal for agricultural use, best used for only limited grazing. This leaves only 259 million acres (105 million ha, or 11 percent) with CLI soil capability classes 1 through 5 suitable for agricultural production. A closer analysis of this potential agricultural resources reveals that only 43 percent of the classified land, or 113 million acres (45.7 million ha), is free from severe environmental restrictions and capable of growing crops. These lands comprise only 5 percent of Canada's land area.

The vast majority of Canada's agricultural land base is concentrated in the prairie provinces (Figure 7). More specifically, Alberta, Saskatchewan, and Manitoba contain 79.2 million acres (32.1 million ha) of CLI class 1 through 3 lands, approximately 70 percent of Canadian total active and potential cropland. The continental climate of the prairies, however, severely restricts the range of agricultural production. Faced with severe winter temperatures, low moisture availability, and short growing seasons, the

FIGURE 7 CANADA'S POTENTIAL CROPLAND (Map courtesy of Environment Canada, reproduced by permission; Simpson-Lewis 1979).

inherently rich lands of the prairies cannot be used for commercial production of many important fruits and vegetables.

Beyond the prairie provinces, the major concentration of agricultural land — almost 16 percent or 17.7 million acres (7.2 million ha) — is in Ontario (Figure 8). Southwestern Ontario represents the largest concentration of prime agricultural land in eastern Canada. Endowed with a favorable climate and situated near urban markets, southern Ontario farmers produce a variety of soft fruits, vegetables, berry crops, special grain crops, as well as livestock and dairy products.

In recent years, the Canadian agricultural land base has been undergoing dramatic change, with two significant trends. The first of these changes is a decline in the quantity and quality of unimproved land. Unimproved lands contain soils with the potential for intensive agricultural use; however, they require corrective measures to reduce limitations or hazards (Canada, Department of Regional Economic Expansion 1965:5). In the past, unimproved land has served as the major reserve for new cropland. Therefore, a decrease in unimproved land reduces the replacement potential for new cropland. In 1921, Canada's farmland was evenly divided between improved and unimproved land, but the 1976 census found 65 percent of the agricultural land was improved with unimproved land down to 35 percent (Statistics Canada 1977).

A second trend has been the westward shift of the land resource base. This transfer has resulted from the increasing loss of agricultural land in Eastern and Central Canada, due to the retirement of marginal land and the conversion of higher quality land near urban areas. Agricultural land is being added in the Prairies and the northern frontier. In Ontario, for example, the amount of agricultural land has shrunk steadily from 1921 to 1976, declining by nearly 20 percent. In sharp contrast, total farmland in Alberta grew by 86 percent during this same period.

Even more serious than the shift in land resources is the quality of land being lost versus the quality of replacement land. Much of the eastern land being converted is prime or high quality land capable of growing fruit, vegetables, or grain, whereas the new western cropland is climatically suitable only for grain (Geno and Geno 1976). The new agricultural districts of Western Canada are especially vulnerable to climatic variability and thus pose high economic risks.

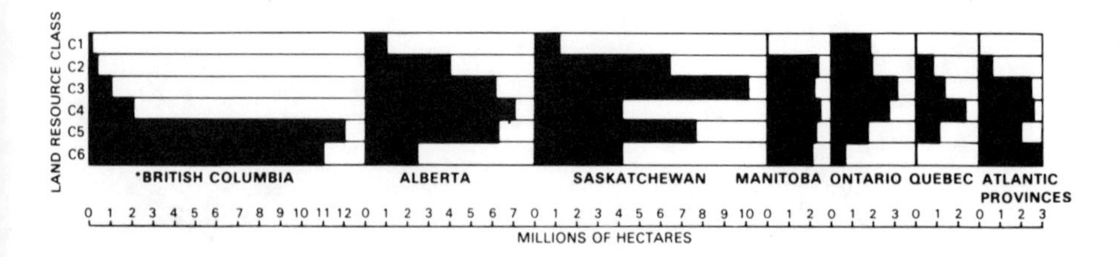

FIGURE 8 TOTAL CANADIAN AREA IN LAND RESOURCE CLASSES (Figure courtesy of J. L. Nowland and G. D. V. Williams, Agriculture Canada. Figure reproduced from *Canada's Special Resource Lands;* Simpson-Lewis 1979). For definition of Land Resource Classes see Table 1, p. 20. Data for British Columbia are estimates.

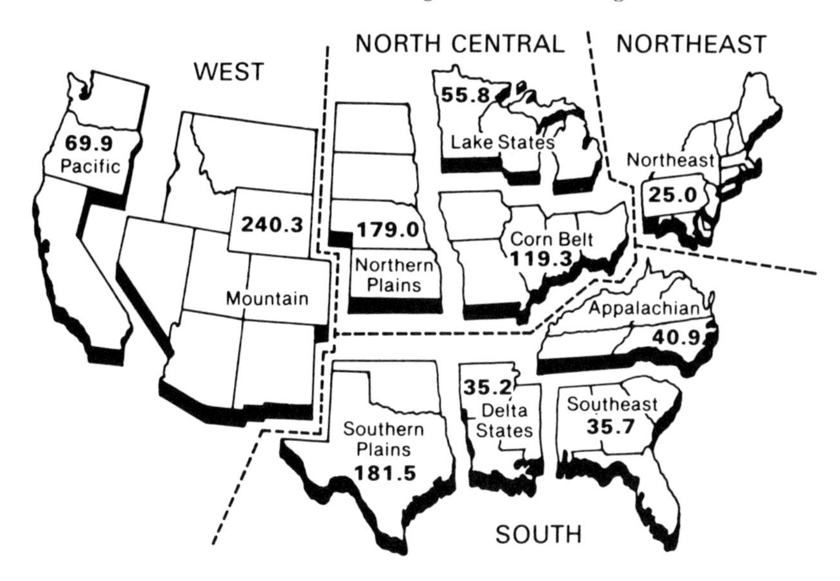

FIGURE 9 TOTAL U.S. AGRICULTURAL LAND USES (1977). Data are in million acres (U.S. S.C.S. 1979).

U.S. Agricultural Lands

The United States has a total of 2.26 billion acres (916 million ha) of land, of which about 1.3 billion privately owned acres (525 million ha) were classified in 1977 by the U.S. Department of Agriculture as agricultural land (NALS 1980). A breakdown into specific land uses reveals about 414 million acres (167.2 million ha) of rangeland; 413 million acres (166.8 million ha) of croplands; 376 million (151.9 million ha) of forestland; 133 million (53.7 million ha) of pastureland; and 23 million acres (9.2 million ha) of farmsteads and "other land in farms."

Within the cropland category, approximately 308 million acres (124.7 million ha) are used for row or closely grown field crops. Orchards, vineyards, and fruit production account for 6 million acres (2.4 million ha), with another 7 million acres (2.8 million ha) in other crops. In any given year 29 million acres (11.7 million ha) are kept fallow. The remaining cropland (190 million acres or 76.9 million ha) is used for pasture, range, or forest.

Geographically, America's agricultural land base is spread throughout the nation (Figure 9). The pattern of current agricultural land uses, excluding the forestland component, indicates smaller amounts of farmland in the more urbanized and densely settled portions of the U.S. Additionally, the large amounts of agricultural land in the Mountain and Southern Plains regions result primarily from rangeland, a very low-intensity agricultural use.

Perhaps a more valuable representation of American agricultural resources is one reflecting the distribution of prime agricultural land. A 1977 survey showed that there are 344.5 million acres (139.5 million ha) of prime farmland in the U.S. (U.S. Soil Conservation Service 1977). Figure 10 presents the regional distribution of this very best agricultural land. The highest concentration of prime land, not surprisingly, occurs

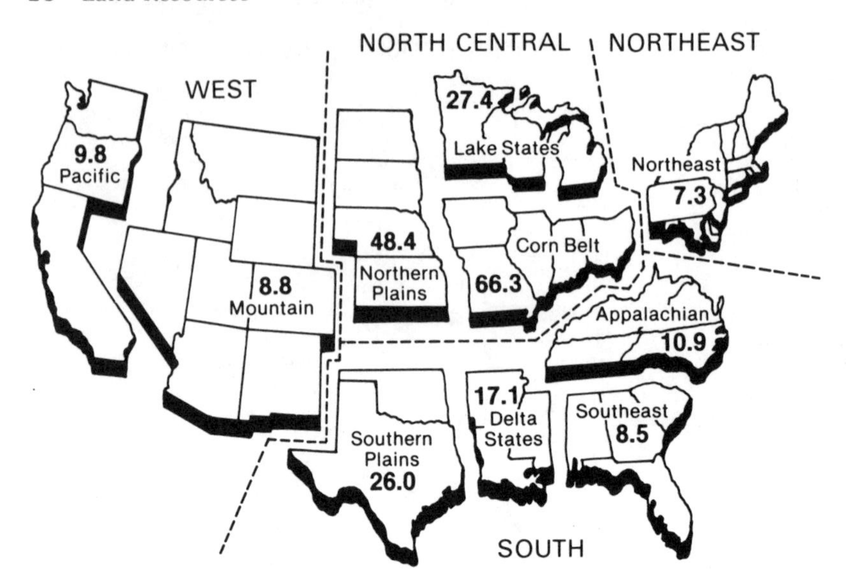

FIGURE 10 PRIME U.S. FARMLAND IN CROPLAND USE (1977). Data are in million acres (U.S. S.C.S. 1979).

in the Corn Belt, followed closely by the Northern Plains. For individual states, the amount of prime land in cropland use ranges from a high of 19.1 million acres (7.7 million ha) in Illinois to a low of 20,000 acres (8,000 ha) in Rhode Island.

In addition to the existing agricultural base, the U.S. Soil Conservation Service estimates there is an additional 126 million acres (51 million ha) of undeveloped land with high and medium potential for conversion to cropland. This new cropland would be added at the expense of other agricultural land uses and a small reserve of idle land. Most gains would also involve environmental and economic costs. According to the SCS, only 36 million acres (14.6 million ha) could be converted simply by beginning tillage. Converting another 44 million acres (17.8 million ha) would involve potential wind and water erosion hazards, but these problems could be handled with well-designed conservation practices. Significant financial investments would be required to bring the remaining 57 million acres (23.1 million ha) into production.

The regions which have the most land for potential conversion are the Southeast, 18.4 million acres (7.5 million ha) with high or medium potential; the Northern Plains, 17.5 million acres (7.1 million ha); and the Corn Belt, 14.5 million acres (5.9 million ha). Relatively little potential new cropland remains in the Northeast and Pacific regions.

Implications for the Future

At a time of escalating pressure on North America's food producing resources, the importance of identifying and managing high quality agricultural land has never been greater. Conversion and loss of agricultural land, particularly prime farmland, to expanding urban and other irreversible land uses has significantly diminished the cropland base and affected environmental quality in both the U.S. and Canada (U.S.

Environmental Protection Agency 1978b; Environment Canada, Lands Directorate 1980).

While neither nation is in danger of "running out" of agricultural land, the cumulative effects of agricultural land conversion pose serious questions for the future. A growing body of evidence indicates that the impressive gains in agricultural productivity resulting from hybridization and greater application of fertilizers may be falling off (Skolds and Penn 1977). With less "prime" quality agricultural land available, greater reliance on marginally productive land will occur, resulting in increased soil erosion, increased fertilizer requirements, and increased environmental damage. Farming lower quality agricultural lands is also more energy intensive.

The economic and environmental costs of overcoming adverse soil and climatic conditions are rising. If these costs are passed on to consumers, the trade-offs surrounding the utilization of land resources will receive more consideration and debate at all levels of government.

4

Agricultural Land Conversion

Unlike population and economic change, the spatial spread of cities has escaped regular and systematic enumeration at a national level. Neither the Canadian nor the U.S. census provides a comprehensive measurement of the spatial expression of urban growth and the loss of agricultural land. Despite the measurement of agricultural area on a regular basis, little is revealed about the direct and indirect loss of agricultural land to urban expansion. Our knowledge of the quality, quantity, and intensity of urbanized agricultural land is based on a relatively small number of public and private studies whose accuracy is limited by the scale of the data base and the size of sample and whose comparability is complicated by variations in frame of reference, definitions, and coverage. Not unexpectedly, therefore, estimates of farmland consumed, intensities of use, and capability of land lost vary considerably, as do projections of farmland loss over the short and long term.

Bearing in mind these measurement and comparability problems, we summarize and analyze the national rate of loss of agricultural land, regional expressions of this process, and some contributing factors. The results from this survey provide a basis for an evaluation of the adequacy of the land resource in relation to current rates of use. We conclude with a consideration of some of the indirect effects of urbanization upon agricultural land.

Land Conversion in the United States

One of the first comprehensive inventories of all major land uses in the United States was undertaken by the Soil Conservation Service in its Conservation Needs Inventory (U.S. Soil Conservation Service 1971). Based upon a two percent random sample of nonfederal land in each country for the years 1958 and 1967, the CNI revealed that in 1967 urban and built-up land (in parcels more than 10 acres in size) occupied approximately 60.9 million acres (24.7 million ha) representing 2.7 percent of the country's land area. During the eight year study interval rural land was consumed at a yearly average of 1.1 million acres (0.45 million ha), enlarging the original urbanized area by 18 per cent (Table 3). The Regional Science Research Institute (RSRI) has estimated that close to half of this urbanized land was originally farmland and one-third cropland (Coughlin *et al.* 1977).

The CNI estimates of rural-to-urban land conversion contrast sharply with those made for the next eight years in the Potential Cropland Study (PCS) of the U.S. Soil Conservation Service (1977). The SCS sampled major land uses in 4554 parcels of approximately 160 acres for 1967 and 1975. Despite the overlap in sample plots between the CNI and PCS studies, PCS estimates that over 2 million acres a year were

TABLE 3 ESTIMATES OF THE CONVERSION OF RURAL LAND TO URBAN AND BUILT-UP USES IN THE UNITED STATES

Studies and Time Interval	Annual Change To Urban And Built-Up Uses[a]			
	All Rural Land	Farmland[b]	Cropland[c]	Prime Land[d]
Soil Conservation Service				
National Resource Inventory 1967-1977 (*after* Plaut 1980)	2,500(100)[e]	1,200(49)	730(39)	910(36)
Potential Cropland Study 1967-1975 (U.S. S.C.S. 1977)	2,079(100)	1,007(48)	606(29)	762(36)
Conservation Needs Inventory 1958-1967 (Coughlin *et al.* 1977)	1,148(100)	563(49)	383(33)	343(30)
Regional Science Research Institute				
(RSRI) (Coughlin *et al.* 1977) 1959-1969	902(100)	442(49)	301(33)	270(30)

[a]Measured in thousands of acres. Urban and built-up uses include residences, commerce, industry, recreation, and transportation

[b]Farmland includes cropland, pasture, and range.

[c]Cropland includes all land used to produce harvested crops. Includes land in rotation.

[d]Includes soil capability classes I and II.

[e]Figures in parenthesis indicate percentage of total land converted.

converted from rural to urban use (Table 3). However, the relative shares that farmland, cropland, and prime land represent in the total converted remained the same. Hence, 36 percent of the land lost to urbanization is prime farmland (Soil Capability Class 1 and 2), representing a disproportionate consumption of high capability land. Some 25 percent of non-federal rural land (excluding urban, rural transportation, and water) is prime land. Vining and colleagues (1977) have shown that 20 percent of the coterminous U.S. but 28.4 percent of its prime land lies within a 50 mile radius of the 100 largest urbanized areas. Even more significantly, counties adjacent to SMSAs which have witnessed the greatest rates of population growth in recent years contain 26.5 percent of the nation's land but 31.5 percent of Class 1 and 2 land. The loss of prime agricultural land becomes more pronounced if we include loss to water, strip-mining, and sterilization because of urbanization. "The U.S. could lose, then, 12 percent of its prime farmland not now in cropland but with medium to high potential for conversion to cropland" (Vining *et al.* 1977:151).

This loss becomes all the more plausible according to the most recent estimates of agricultural land conversion by the SCS through the National Resource Inventory (U.S. Soil Conservation Service 1979). With a sample size five times that of the Potential Cropland Study, results from the NRI indicate a continued expansion in the magnitude of farmland loss over the original CNI estimates. Whereas urban and built-up areas equalled 61 million acres (24.7 million ha) in 1967, by 1977 they had grown to 90 million acres (36.4 million ha; Dideriksen *et al.* 1979:21). This 47.5 percent increase in built-up area represents an average annual increase of 2.9 million acres (1.2 million ha) per year — nearly three times the rate of the 1958-1967 period. Plaut (1980) has adjusted these estimates downwards because of evidence that the 1967 base year data ignored some existing urban development (Table 3). Despite this adjustment, the NRI estimates of annual rates of rural to urban land conversion are also significantly higher than the Potential Cropland Study estimates. Certainly much variation can be traced to the larger sample size of the NRI study.

The National Agricultural Lands Study (NALS 1981c) also relied upon SCS's Potential Cropland Study and the National Resource Inventory for its analysis of the conversion of agricultural land (Table 4). A cooperative effort by the U.S. Department of Agriculture and the Council on Environmental Quality to examine the changing agricultural land use base within the nation, the NALS aggregated all major sources of agricultural land conversion. NALS derived an annual average rate of conversion of 2.9 million acres (1.2 million ha), 70 percent of which was consumed by urban, built-up, and transportation uses and 30 percent by man-made water impoundments. Of all land converted, 23 percent was classed as cropland. The rate of conversion is further increased if the movement of land into speculative and development uses is added with urban and water uses. This brings the annual rate of conversion to 5.5 million acres (2.2 million ha).

Regionaly there are significant variations in the changes to the agricultural land base (Figure 9). In the Northeast, which contains 25 percent of the nation's population and 4 percent of the cropland, urban, built-up, transportation, and water uses accounted for a loss of 3 million acres (1.2 million ha). Roughly 20 percent of this loss was cropland, a proportion equivalent to the ratio of cropland to total rural land uses. The South accounted for over half of all the land converted to non-agricultural uses for the same period. At the same time approximately 15 million acres (6.1 million ha) shifted out of crop production into forestry, partly in response to increasing importance of forestry in the Southern economy but also because of declining soil fertility. Despite high population growth, the Potential Cropland Study revealed that a greater proportion of Southern cropland was converted to water than to urban or urban-related uses.

In contrast with the South, the North Central region is not only the most industrialized region of the U.S., but also the region with highest concentration of prime farmland (Vining *et al.* 1977:148). In ranked second of the four census divisions in terms of agricultural land lost but first in terms of proportion of cropland converted (31 percent). NALS argued that the greatest regional threat to cropland is from stripmining,

TABLE 4 AGRICULTURAL LAND CONVERTED TO URBAN, BUILT-UP, TRANSPORTATION, AND WATER USES, BY FORMER AGRICULTURAL USES, 1967-1975[a]

	Cropland					
Census Region	Cropland's Share of Region's Rural Land 1977 (%)	Quantity of Cropland Consumed (10^6 acres)	Cropland's Share of Rural Land Converted (%)	Pastureland and Rangeland Consumed[b] (10^6 acres)	Other Agricultural Uses Consumed[c] (10^6 acres)	Total
Northeast	20	0.6	20	0.1	2.3	3.0
South	22	2.5	21	2.1	7.4	12.0
North Central	35	1.6	31	0.8	2.8	5.2
West	18	0.7	23	1.3	1.0	3.0
Total	37	5.4	23	4.3	13.5	23.2

[a]Measured in millions of acres. Adapted from NALS Final Report 1981

[b]Land producing forage plants for animals.

[c]Includes forestland, water, land reserved for wildlife, commercial feedlots, greenhouses, and nurseries.

which promises to expand in this area. The last census region, the West, lost the second highest proportion of cropland (23%) to urbanization and water projects. This magnitude should not be surprising since California represents the main pressure point for agricultural land conversion in the nation — a function of strong population growth (5 million persons between 1970 and 1975) as well as the juxtaposition of urban and food-producing areas. While both the West and North Central regions experienced shifts from other rural uses into cropland, offsetting the impact of urbanization pressures,the pressure on water resources from high rates of population growth is likely to slow that shift in the future.

To date there is no satisfactory explanation for the marked divergence between the Potential Cropland Study and Conservation Needs Inventory estimates of farmland conversion. Since the CNI classified urban areas under 10 acres in size as "other land," rapid growth in these small areas may have changed their status and classification to urban and built-up uses (Lee 1978; Plaut 1980). For example, the PCS found that 25 percent of the rural to urban land converted for the period 1967 to 1975 was formerly "other land" (Plaut 1980:538). This factor alone could account for a significant proportion of the variation in estimates in rates of land conversion between the two studies.

In a more systematic attempt to account for recent estimates of land conversion, Vining and associates (1979) first examined earlier area and population relations. Specifically, they regressed percentage of a county's land area that is built up against its population density for the years 1958 and 1967. Using a curvilinear relationship, the authors found the exponent to be unchanged but the coefficient increased substantially between the two years. In two separate estimates of urban land use needs — one assuming a stable function over time, the other an increasing constant — they could account for the dramatic shift in land consumption between the periods 1958 to 1967 and 1967 to 1975. They concluded that "the rate of growth in the urban land estimated by the Potential Cropland Study for the period, 1967-1975, cannot be accounted for on the basis of the relationship between population and land area that held true in the past or even on the basis of changes in these relationships that were observed in the past" (Vining *et al.* 1979:153). Instead, they explained the high rate of land conversion in terms of the lowering of development densities for "urban purposes at all densities."

Where Vining and colleagues accepted the SCS estimates and attempted to explain them with area and population relationships, RSRI (Coughlin *et al.* 1977) questioned the validity of the SCS estimates and attempted to generate its own on the basis of the relationship that Vining and associates found so unsatisfactory. Briefly, RSRI regressed percent of county land in urban areas against housing density for 1960. The resulting regression parameters were then multiplied by yearly housing starts for the years 1960 to 1970. The results, summarized in Table 3, are considerably lower than those of CNI for the years 1958-1965.

At this stage more can be learned about national rates and certainly more about development densities if we turn our attention to case studies of the rates of rural to urban land conversion.

Selected U.S. Case Studies

Most of what we know of rural to urban land conversion comes from studies of separate regions of the country. Units of analysis range from urbanized areas depicting the geographical extent of urban growth and SMSAs representing the functional urban region to the much more arbitrary counties. Table 5 provides highlights from the

TABLE 5 CASE STUDIES OF RURAL-TO-URBAN LAND CONVERSION

Study Areas and Time Intervals	Population and Area			Intensity of Use		Land Quality	
	Share of Change in National Population (%)	Rate of Change in Population (%)	Rate of Change in Urban Area (%)	Acres Per 1000 Persons (Base Year)	Acres Converted Per 1000 Increase in Population	Cropland as a proportion of all rural land	Cropland as a proportion of rural land consumed
United States							
96 Counties in 12 Northeastern States 1950-1960 (Dill & Otte 1971)	17.4	16.7	—	204	220	24	50
RSRI Estimates for Massachusetts 1951-1971 (Coughlin *et al.* 1977)	—	—	—	—	280	9.1	15
157 Urbanized Areas							
1950-1960	75.3	30.1	76.6	185	469	—	—
1960-1970 (Hart 1976)	74.6	19.6	35.1	251	448	—	—
53 Rapid Growth Counties 1961-1970 (Zeimetz *et al.* 1976)	20	46.2	27.1	233	173	32.9	34.6
Canada							
Environment Canada (Gierman 1977) 71 Centers, 1966-1971	79.7	10.1	17.1	101	172	45	62.9
Environment Canada (Warren & Rump 1981) 80 Centers, 1971-1976	60.2	6.1	10.4	106	178	45	61.3
4 Centers, 1951-1960 (Gertler & Hind-Smith 1961)	2.2	52.8	66.9	105	103	—	62

findings of seven studies into the relationship between urban growth and loss of rural land in the United States and Canada.

Not unexpectedly there are significant variations among proportions of cropland converted, intensities of development, and growth in urban areas. For 96 counties within the Northeastern United States, Dill and Otte (1971) found an average of 220 acres converted per 1000 population increase. Although this figure conforms closely to existing densities, the proportion of cropland converted was more than double its share of rural land use. The RSRI (Coughlin *et al.* 1977) estimates for Massachusetts (based on Foster 1976) also show a disproportionate loss of cropland. RSRI's estimate of development density parallels Bogue (1956), who derived values ranging from 172 to 262 acres per 1000 population increase for 147 SMSAs between the years 1930 to 1950. Contrasting with these localized studies are the analyses of Hart and Zeimetz and colleagues. In Hart's (1976) study the original (1950) urbanized areas experienced high per capita rates of land consumption for both time periods. Hart did not address the sizeable difference between existing densities and the intensity of new development. Since urbanized areas include settlements with densities as low as 1000 persons per square mile (or 640 acres per 1000 persons), it is not surprising that estimates of per capita development are high. At the other end of the size spectrum, pressures for space are clearly dictating more efficient development densities. For example, the 20 largest urbanized centers for the U.S. converted rural land (1960 to 1970) at an average rate of 358 acres per 1000 population increase. Whatever group of cities one uses, however, urban area grew at a faster rate than population for both periods.

In Zeimetz's study (Zeimetz *et al.* 1976) the reverse was the case (Table 6). The use of a large systematic sample of land use in 53 rapid growth counties permitted more precise delineation of land uses and, hence, the allocation of land to urban uses. These counties accounted for 20 percent of the gain in U.S. population between 1960 and 1970. For the study area as a whole, 770 thousand acres (312 thousand ha) of rural land were converted, 35 percent of which was cropland. The proportion of cropland in urban development varied significantly among the study regions with proportions ranging from highs of 70 and 62 percent in California and the Great Lakes, respectively, to lows of 19 and 6 percent in the Piedmont and Florida Gulf. Generally, this variation in the contribution of cropland to new urban development reflects the spatial variation in the availability and importance of cropland across the country. Coughlin (1979:43) has argued that urban development is no more biased toward cropland than to other relatively flat open land. In fact pasture, range, and open idle land have the greatest likelihood of being developed, often by a margin of 2 to 1.

Idle land represented the second most important source of land for urban development (Zeimetz *et al.* 1976). It represented only 9.5 percent of the study area and yet an average of 33 percent of urban development land. Clearly this proportion reflects the comparative advantages, both economic and political, of developing cleared but unused land. That much of this idle land had high capability is indicated by the fact that approximately 45 percent of the gross decline in the cropland base was due to idling (compared to 33 percent due to urbanization). Some 64 percent of the gross additions to idle land came from cropland.

How efficiently or intensively was this land developed? Table 6 illustrates that with every 1000 increase in population an average of 173 acres (70 ha) of rural land was withdrawn from production (5.8 persons per acre). For residential land itself this figure declines to 106 acres (43 ha). These figures conceal dramatic variations in development densities. California was the most efficient user of land (97 acres or 39 ha per

TABLE 6 PER CAPITA CONSUMPTION OF RURAL LAND IN 53 FAST-GROWTH COUNTIES

	Acres Converted Per 1000 Population Increase	
Region	All Urban Uses	Residential Uses
Northeast	181	109
Middle Atlantic	137	95
Piedmont	216	141
Appalachian Fringe	275	154
Florida Gulf	481	341
Corn Belt	142	92
Great Lakes	173	75
S. Central Prairie/Woodland Fringe	146	117
Texas Prairie	202	92
Colorado	234	131
California	97	57
53 County Total	173	106

Adapted from Zeimetz *et al.* 1976.

1000 population), whereas the Florida Gulf was the least efficient (481 acres or 195 ha). The Great Lakes region was closest to the average.

Land Conversion in Canada

Unlike the United States, Canada has no comprehensive national survey of the quantity, intensity of use, and agricultural capability of rural land converted to urban, built-up, transportation, and water uses. But a useful substitute is Environment Canada's very detailed and systematic land use data collected for the periods 1966 to 1971 and 1971 to 1976 representing all urban areas (Gierman 1977; Warren and Rump 1981). Urban area is all land in Census Metropolitan Areas, Census Agglomerations, or other urban centers over 25,000 persons that is classified as "built-up" by the CLI Land Use Classification (Gierman 1977:71). Centers of this size account for 66 percent of the Canadian population and the majority of population growth. They are generally located in regions with Canada's best cropland.

Unfortunately, little is known of the character of urban expansion in smaller centers. Considering the renewed importance of these centers as sources of decentralized growth, national estimates of the impact of urban growth on agricultural land will be incomplete until they are taken into consideration. Notwithstanding the bias toward larger centers and lack of information on transportation and water resource demands, the quality of the land use data provides a very useful account of current land consumption trends in the urban development of Canada.

Crerar (1960) was the first to measure systematically the loss of agricultural land within the vicinity of metropolitan centers. Using an approach similar to Bogue's (1956), in which change in farmland acreage is the measure of rural to urban land consumption, he found that metropolitan centers were consuming an average of 382 acres (155 ha) per 1000 population increase. Since his sample population lived at an average density of 108 acres (44 ha) per 1000 persons Crerar argued that cities in Canada were wasting 2.54 hectares for every one that they consume.

More precise estimates of farmland loss and urban fields of influence (urban shadow) were provided by Gertler and Hind-Smith (1961). Their estimates of the efficiency of urban development are lower than Crerar's, indicating that a proportion of his estimates of farmland removed from agriculture was not consumed directly by urban growth. However, the correspondence between the two sets of estimates increases when both direct (urbanized) and indirect (idling of farmland) influences are combined. On average, 303 acres (123 ha) were either consumed or affected by urban development for every 1000 increase in population.

The main concern of Environment Canada (Gierman 1977; Warren and Rump 1981) in its two land use studies was to measure the direct effects of urbanization on rural land. Goals included providing an inventory of total land consumed, the capability of the land (for a variety of functions), intensities of use, and regional impacts.

Canada's urban areas of over 25, 000 persons occupy approximately 0.07 percent of the country's total land area. The scale of urban development is therefore considerably smaller than that of the United States. However, given the size of the cropland base (296 million acres; 120 million ha), its agroclimatic limitations, and its proximity to Census Metropolitan Areas (26 percent of all Class 1-3 land lies within a 50 mile radius of CMAs), this figure understates the pressures of urban growth on the land resource base. For example, between 1966 and 1976, 62 percent of all land urbanized in Canada was Class 1-3. A disporportionate share of prime land is consumed because some of the fastest growing centers are surrounded by an above average proportion of cropland. Southern Ontario, with the highest proportion of cropland, also accounted for the greatest rates of land conversion.

Southern Ontario and Western Canada (Manitoba, Saskatchewan, and Alberta) account for approximately three quarters of all urban development. With respect to the rate and magnitude of agricultural land lost to urbanization, significant changes occurred between the two study periods. The most important of these changes are an increase in the proportion of improved land urbanized from 54 to 58 percent, but a 28 percent decrease in the quantity of land urbanized; an increase in per capita rates of land conversion from 172 to 178 acres per 1000 population increase; an increase in the proportion of rural land lost in Western Canada; and an increase in the share of urban development of smaller urban centers (25,000 to 100,000) from 23 to 29 percent. Thus, Canada has experienced increased population growth in smaller urban centers (as in the U.S.), decreased land use pressures overall because of a decline in the rate of increase in population growth, and a western shift in population growth. These factors would explain the lowering of population densities, the increased share of urban physical development by smaller centers, and the greater share of urban development in Western Canada.

Great spatial variation in the per capita rates of land conversion in both study periods is largely explained in terms of city size and rates of population growth. Cities ranging in size from 25,000 to 50,000 consumed an average of 469 acres (190 ha) per thousand population increase between 1971-1976 compared with the largest centers (500,000+) which consumed land at an average density of 121 acres (49 ha) per 1000 population increase. These values closely parallel those of Ziemetz and associates (1976) for fast growth regions in the U.S. Unlike the U.S., urban land area grew at a faster rate than population in both Environment Canada's studies.

Although systematic analyses of the Environment Canada data are rare, one study revealed some interesting correlates of change in the land conversion process (Pierce 1981). Whereas change in population accounted for much of the variation in the rate of rural to urban land conversion, other significant factors appeared to be the economic structure of the urban region and the capability of the land for agriculture. For example, when the influence of population change and soil are adjusted for, extraction and transportation centres showed themselves to be very space-extensive. A weak inverse relationship was found to exist between cropland's share of total land converted (class 1-3) and per capita rates of land conversion. Bryant (1976), studying farm generated determinants of land use change in the urban fringe of CMAs in Canada, found significant differences in the response of the agricultural sector to urbanization pressure. These differences were attributed to the quality of the resource base (more prosperous regions were better able to withstand urbanization pressures) and size of city (high absolute levels of development regardless of growth led to degeneration in agriculture).

Supply-Demand Projections of Agricultural Land Adequacy

Estimates of the quantity of agricultural land lost to urban expansion are, in themselves, meaningless unless placed within a wider context of future demand for food and the supply of land required to meet that demand. If some uncertainty surrounded estimates of farmland loss, even greater uncertainty surrounds the estimates of land resource requirements for the future. Assumptions must be made about rates of population growth, changes in per capita income, changes in demand and its impact spatially, the real price of food, and the relative importance of land in the production process. Recognizing the complexity of the task and the limitations inherent in most projections, we briefly examine studies of the main supply-demand relationships as they relate to the United States and Canada.

United States Projections

Projections of the urbanization of agricultural land are usually based upon a simple linear extrapolation. Plaut (1980), for example, assumed a cropland loss of 730,000 acres a year until the year 2000. With a reserve in 1977 of 135 million acres (55 million ha) this rate of growth would consume an area equal to 12.4 percent of the reserve and 44.5 percent of prime agricultural land. Hart (1976) in his study of urban encroachment on rural areas estimated the loss of rural land to urban and built up uses until the year 2000 to vary from 32 million to 38 million acres (13-15 million ha). Assuming that cropland will continue to represent a 30 percent share of rural land converted, between 9.6 and 11.4 million acres (3.9-4.6 million ha) will be retired from cultivation, or between 7 and 8.4 percent of the reserve. The difference between the two sets of estimates reflects, at least in part, Plaut's use of the NRI data and Hart's use of the much more conservative CNI data.

If, for the moment, we accept these estimates as being both possible and plausible what are the projections for changes in demand and how will these changes relate to the quantity of cropland needed to meet this demand? The NALS, using growth scenarios developed by USDA, estimated that the volume of demand for U.S. agricultural products is expected to increase between 60 to 85 percent above 1980 levels over the next 20 years. This prediction assumes constant real prices and annual growth rates in demand of between 2.25 and 3.1 percent.

Additional output to meet the burgeoning demand can be generated from increasing productivity, an increase in land area, or both. The first involves the use of land- and labor-augmenting technologies which are energy intensive and associated with hybrid varieties, pesticides, fertilizers, farm machinery, and irrigation. Throughout the 1950s and 1960s the substitution of these inputs for land contributed to an annual average increase in total productivity of 1.9 percent on a declining agricultural land resource base (Crosson 1977:52). The trend toward energy-intensive production was largely the result of the slower rate of price increases for nonland versus land inputs. Crosson (1977; 1979) argued that the price conditions which encouraged the previous type of factor substitution have now changed to encourage land-using technologies. But price alone cannot explain this shift. Evidence of a decline in the rate of increase in productivity of major crops during the first half of the 1970s points to a decline in the marginal productivity of fertilizers and possibly other energy-intensive technologies. According to Crosson, both conditions, high prices for non-land inputs and declining marginal productivity, will lead to greater demands for land resources. It is certainly unlikely that during the next ten years we will be able to return to annual average rates of increase of productivity of 2 percent and above.

All of the productivity increase estimates used by NALS were between 0.75 and 1.5 percent. Using this range of estimates, and assuming that total demand will grow by an annual rate of 2.75 percent for the period 1980-84 and then by 3.10, 2.60 and 2.25 for the three succeeding quinquennial periods, NALS then projected required cropland acreage (Table 7). Clearly under conditions of high demand (3.10%) and low productivity gains (0.75%), acreage requirements are high. Between 77 and 113 million additional acres will be needed to meet projected conditions by the year 2000. Plaut (1980) has estimated a minimum of 79 million additional acres (32 million ha) by the year 2000 under what he terms pessimistic conditions — rising real prices and average annual gains in productivity of 1 percent and in demand of 2.77 percent.

If we accept Plaut's projection of the level of urbanization of cropland (14.6 million acres or 5.9 million ha) to the year 2000 and the NALS assumption of high demand for cropland and low increases in crop productivity then projected additional urban and agricultural acreage requirements by the end of the century could range from 92 to 128 million acres (37-52 million ha). Given that these estimates ignore the impact of other non-agricultural demands on the cropland base, these figures could be conservative. Exactly where will this land come from? If we assume that there will not be major changes in dietary habits, (a shift from foods which are indirectly produced by cropland, such as beef, to those directly produced, such as soybeans), then we can expect that a greater proportion of the existing cropland base will be used. But this must be made up from the potential cropland base. In short, we are dealing with a zero sum situation. Recall that SCS estimated there were some 127 million acres (51 million ha) with high to medium potential for conversion to cropland. While it is, of course, technically feasible to convert this land, a number of economic obstacles will contribute in the long run to

TABLE 7 PLANTED ACREAGE NEEDED TO MEET PROJECTED U.S. DEMAND FOR PRINCIPAL CROPS BASED ON THREE RATES OF CROP YIELD GAINS, 1980-2000 (million acres)[a]

Cases and Annual Yield Gains	1980	1984	1989	1994	1999
Case A: .75% gain in Crop Yields	294	325	365	400	407
Case B: 1.25% gain in Crop Yields	294	317	346	370	389
Case C: 1.5% gain in Crop Yields	294	313	339	358	371

[a]Projected demand levels assume constant real prices. Principle crops include about 90 percent of the total historic planted acreage for all crops.

Source: National Agricultural Lands Study 1981c.

higher land costs, crop prices, and numerous environmental costs. Some 36 million acres (14.6 million ha) of high potential land can be converted at relatively low cost but the remaining 91 million acres (36.8 million ha) of medium potential land will require more effort and are subject to greater erosion. Since a relatively high proportion (44.5%) of the prime or high-potential cropland will be subject to urbanization pressure during the next 20 years, the demand for, and costs of, conversion will be even higher for the medium-potential land.

Obviously the accuracy of these estimates is dependent upon the validity of the assumptions. If the assumptions are valid, then by the year 2020 a major supply-demand imbalance will have developed. Urban centers will no longer have abundant rural land at their disposal, nor will farmers be able to shift easily to high or medium potential cropland. Critical here is not that 4 percent of the U.S. land mass will be urbanized in 20 years time as Hart (1976) seems to emphasize, but that the equivalent of 10 percent of the potential cropland and half of the potential prime land currently in reserve will be converted to urban uses. That urban growth is consuming a disportionate share of the cropland base and that most, if not all, of the reserve will be needed in the future points to the need for improvement in the allocation of rural land.

Canadian Projections

Canada possesses no major reserves of prime agricultural land. Potential cropland exists in Western Canada (most notably in the Peace River area of British Columbia and Alberta) but in terms of class 1 equivalents or in terms of its agroclimatic rating it is far inferior to the cropland currently undergoing urbanization. Again, class 1 equivalents refer to a productivity rating (Hoffman 1976) in which class 2 land is 80 percent as productive as class 1; class 3, 64 percent; and class 4 only 49 percent as productive as class 1 land. From a different perspective, the Agroclimatic Resource Index (ACRI; Williams 1974) provides a ratio of the number of forest free days required for barley to mature (60 days) adjusted for moisture constraints. Most of the potential cropland is class 3 and 4 land and is characterized by an Agroclimatic Resource Index of 1.0 to 1.5. Williams and colleagues (1978) have estimated that 50 percent of the country's population live in regions which contain the best 5 percent of the country's agroclimatic land resource (ACRI = 2 to 3). As we have already observed, approximately one-quarter of all class 1-3 land in Canada lies within a 50 mile radius of its CMAs.

Williams (1973) estimated that between 1961 and 1971 three quarters of all urban growth took place on the agroclimatically rated best 5 percent of farmland. If we assume that an additional 11 million urbanites will be living in Canada by the turn of the century and if we also assume development will occur at 100 acres (40 ha) per 1000 population increase (a conservative estimate), then 1.1 million acres (0.45 million ha) of rural land will be converted to urban uses. If urban development displays the same preference for prime land as it has in the past, then nearly 825,000 acres (334,000 ha) or 10 percent of Canada's best agroclimatic land resource will be urbanized (Williams *et al.* 1978). If new development occurs at more realistic densities of 200 acres per 1000 population increase then 20 percent of this agroclimatic land resource will disappear. Without dramatic increases in productivity and with marginal reserves, Canada is likely to face a net decline in both its productive base and its productive potential by the year 2000.

Other Impacts of Urbanization

Aside from the actual loss of agricultural land to urban development, how else might this land resource base be affected by urbanization? We know, for example, that the idling of farmland, changes in the structure of farming, and declining farm infrastructure are often indirect effects of urbanization or, as we suggested in Chapter 2, intervening forces of change between the growth and distribution of population on the one hand and, on the other, the attitudes and behavior of farmers toward production decisions. Not satisfied with establishing the existence of these intervening forces, many researchers have turned their attention to the variation within and between regions in the effect of urbanization pressures on agriculture's land resource base.

Land use change in the urban fringe and shadow is the result of a complex interplay of farm, urban, and societal factors (Bryant 1976). As Bryant and Russwurm (1981:34) have observed, "agricultural change in urban regions is seen increasingly as the product of urbanization processes and agricultural processes, both of which may exhibit significant regional variations."

Bryant (1976) tried to unravel some of the complexities involved in the interaction between urbanization and agriculture in his analysis of regional variation in agricultural land use change between 1961-1971 in the vicinities of Canadian CMAs. Arguing that land use change is a function of not just urban based forces but also of the physical environment for and the structure of farming, he created four types of regional environments on the basis of urbanization presssures, structure of farming, viability of agriculture, land resource base, and barriers to expansion. The regional environments were then used to account for land use change in CMA based regions. Bryant found that relatively slow-growing regions experienced some of the largest proportional decreases in total farm acreage, fast-growing regions some of the smallest. In fact, a few CMAs from the latter group experienced an increase in improved acreage. Economic obsolescence and uncompetitive agriculture explained the demise of agriculture in the slow growth region. Slowing the rate of loss in fast growth regions was an increase in part time and hobby farming.

Absolute size of an urban region, as opposed to rate of growth, appears to be significant in affecting the type of agricultural fringe that develops. Bryant concludes with the observation that in addition to agriculture's responses to urbanization and to 'marginality,' a third type of change exists, a combination of the previous two which he terms 'normal' change.

In the United States, Berry (1978) examined idle land as a proportion of total farmland according to the soil capability of the land and the population density in 1970. He found a statistically significant decrease in idling with greater soil capability in areas of low density, suggesting that response to marginality was an important factor in the proportion of idled land. Hart's study of the abandonment of cleared land in the eastern United States (1968) supports this finding. However, in the higher density areas no significant decrease emerged. Idling due to urbanization pressures accounted for approximately 35 percent of the farmland within the higher density areas.

Perhaps less obvious than idling but nevertheless a significant feature of the interplay between agriculture and urban areas is the changing structure of farming. Earlier we offered some possible explanations for changing structure of farming in the vicinity of urban areas. Uncertainty created by urbanization pressures diminished the incentive for investment in certain types of agriculture. The time left for agriculture, then, was a critical consideration in farming operations. For example, Berry found in a study of changes in types of farming in both metropolitan and nonmetropolitan counties in New Jersey, New York and Pennsylvania that "dairying and related activities are most sensitive to urbanization and population growth and are the kinds of activities which decline most consistently in metropolitan counties and rapidly growing nonmetropolitan counties" (Berry 1978:5). High levels of investment required to initiate and sustain milk production discourage dairying in areas threatened by urbanization. Interestingly, activities not requiring significant investments or planning horizons, such as corn and alfalfa production, did not decline in fast-growing counties during the early 1970s. Improved prices during this period may have strengthened these activities, suggesting that the right economic climate for farm production can offset, at least over the short term, urbanization pressures.

Walker (1981), in a behavioral profile of farming in Toronto's urban shadow, found that the farm community was, in general, withstanding urbanization pressures despite a nonfarm population of between 80 and 90 percent. He attributed the stability of the system to the introduction of regional planning which has reduced agricultural uncertainty. "The farmers know which areas are slated for urban development and which are likely to retain their agricultural character" (Walker 1981:197).

Further insight into how the agricultural land resource base responds to urbanization is provided by a recent study into the nature of the land market at the periphery of six metropolitan centers in the U.S. and Canada. The study found that although the majority of parcels in the rural-urban fringe have small acreages, the major portion of the land area was defined by large holdings (Brown *et al.* 1981). Whereas most of the land surrounding the two Canadian cities (Toronto and Calgary) was still farmed (79%), 50 percent of the land area in the U.S. fringe was under residential use. In both the U.S. and Canada, "personal 'users' account for fully three-fourths of all owners in the sample and as much of the land area" (Brown *et al.* 1981:134). But these users are not particularly active in the land market, which contrasts sharply with the developers who, despite their small proportion of parcels and area, actively pursue the development options available to them. When the sample was divided into areas experiencing weak versus intense development pressure, differences emerged in land holding characteristics and in size and use of rural land holdings. Investors and developers own a larger proportion of land holdings as development pressures rise. Moreover, most of the recent buyers in the urban fringe are investors and developers. Also associated with increased pressure are shifts in ownership organization towards partnerships, syndicates, and corporation. As one would expect, agricultural production declines, as does average parcel size as the promise of urbanization increases. Despite the fact that the rate of land transactions accelerates for land owners as development pressure increases, personal factors form an important element in the decision to sell. The study reports that family or life-cycle factors were given as the main reason for wanting to sell for over 33 percent of all respondents in U.S. cities.

What can we conclude about the interplay between urbanization pressure and agriculture's resource base from this sampling of recent research? Variation among cities in land idling can be traced to the rate and existing scale of urbanization pressure, the strength and viability of the agricultural sector, and agricultural commodity prices. Within the urban fringe evidence suggests that a reduction in uncertainty through planning can improve the environment for farm production. But the complexity of land use change within urban areas points to the need for securing not only protection against urbanization, but also the conditions which will encourage a viable agricultural industry.

5

Planning to Protect Farmland

Faced with increasing public concern over agricultural land conversion, the issue of farmland protection has become an important priority for political leaders across North America. Whereas most community leaders would support the basic notion that agricultural land resources are needlessly being wasted, agreement on how to alleviate the problem is far from unanimous. A survey of public policy in Canada and the United States finds a variety of programs and strategies, from rigid police power controls to indirect financial compensation approaches (Furuseth and Pierce 1982). In a few cases, lack of action or policies is still the rule.

Institutional Background

The history of direct public policy affecting agricultural land use is by necessity brief. Generally, most observers suggest that 1956 marks the first large-scale direct policy action in the United States (Nielsen 1979). In Canada discussions surrounding the need to protect farmland date back to the early 1960s (Crerar 1960), but large-scale policies were not enacted until 1973 (Pierce and Furuseth 1982). In the intervening period, public policies have diffused throughout the continent and through all levels of government — local, provincial/state, and federal.

At the local level, municipalities, counties, and regional governments have, over the past decade, become more sensitive to the effects of expansive low-density growth on their community. In many areas planners are now charged to guard against the fragmentation of agricultural resources and leap-frogging urban growth. *Ex post facto* research, however, suggests that most local government actions have not been particularly effective (Esseks 1978; Krueger 1980; People for Open Space 1980; Pryde 1982). Perhaps the failure of locally initiated policies reflects the pluralistic political structure of local government, politics dominated by growth-oriented interest groups. Measures to protect agricultural resources may be enacted, but eventually they are ignored or compromised when they affect the growth of the community (Esseks 1978; Krueger 1980; People for Open Space 1980).

Generally, the most important government bodies have and continue to be provincial and state governments. The dominance of provinces and states on this particular issue is not surprising. These governmental bodies are legally, politically, and institutionally best equipped to respond to broad land use questions. In Canada the British North America Act vests provincial governments with broad powers to regulate land use in the public interest, while in the United States the Federal Constitution provides similar authority to the states. Traditionally, these senior governments have assigned responsibility for land use planning and policy making to regional, municipal, and county governments. However, on land use issues of supra-local importance, or in situations where local decision making has failed to uphold the public interest, provinces and states have reclaimed these powers. Both of these circumstances have contributed to the evolution of provincial and state farmland protection programs. In most cases these policies require some type of local participation. The provincial or state governments establish the rules, scope, and procedures, and then assign the implementation responsibilities to local government (Goodwin and Shepard 1974). Oversight by state or provincial agencies insures that local governments are carrying out their requirements.

At the national scale, the federal governments of both Canada and the U.S. have treated agricultural land use policy as a provincial or state concern. Until 1976 in the U.S. (Knebel 1976) and 1980 in Canada (Environment Canada 1980) government had not directly addressed the question of farmland alienation with any specific policy statement or guide. Both federal governments are now moving toward a more activist stance with the consideration of new policies to facilitate the retention of agricultural resources (Dunford 1982). However, the level of federal involvement is minor when compared to provincial, state, and local initiatives.

A review of the programs currently in operation across North America suggests a typology of farmland protection strategies. Individual programs are classified in Figure 11 based on generic or structural characteristics.

Financial Compensation Strategies

The first group of programs — the financial compensation strategies — are designed to offset the impact of nearby or approaching urban land use on land values. These programs provide direct or indirect financial benefits for the maintenance of agricultural land uses. In a majority of programs a financial subsidy is provided to farmland owners on a regular, usually annual, schedule. The assistance can, however, be in a single larger transfer payment. In either case the expectation is that given financial assistance, farmers will keep their land in agricultural use, foregoing the opportunity to convert their land.

Farmland Protection Typology & Matrix

FARMLAND PROTECTION STRATEGIES	Implementation Scale			Type of Measure		Focus		Likely Effectiveness		
	Local	Provincial/ State	Federal	Incentive	Regulatory	Indirect	Direct	Slight	Moderate	Strong
Financial Compensation Mechanisms										
Use Value Property Taxation	●	●		●		●		●		
Circuit Breaker Taxation		●		●		●		●		
Property Tax Credits		●		●		●		●		
Inheritance/Estate Tax Relief			●	●		●		●		
Transfer of Development Rights	●	●		●			●		●	
Police Power Mechanisms										
Agricultural Zoning	●	●			●		●		●	
Provincial Police Power		●			●		●	●		
Comprehensive Mechanisms										
Integrated Provincial-State Programs		●		●	●		●			●
Agricultural Districting	●	●		●	●		●		●	

FIGURE 11 FARMLAND PROTECTION TYPOLOGY AND MATRIX.

Tax Policies

The primary mechanism used in these programs is tax relief. Since taxes, especially property taxes, constitute a major cost to farmers and can be manipulated to provide benefits for selected classes of taxpayers, their use as a tool to protect farmland is not surprising.

At the present time, both property tax and inheritance or estate tax programs have been modified to provide incentives for agricultural land use. Property taxation strategies have evolved into the most widespread tool for encouraging and assisting agricultural land use in North America (Table 8). The Federal Tax Reform Act of 1976 which reshaped many aspects of federal estate and inheritance taxes in the U.S. has spurred a reduction in death taxes on farmland and property in many states, but Canadian inheritance laws remain generally unchanged.

Traditionally, the property and estate tax laws in North America have relied on an *ad valorem* tax assessment formula. Given the goals of maximizing public revenues, the advantage of this policy is obvious. The tax is proportional to the value of land in its highest and best use. Hence, the value of a property is defined as the price it would sell for if market or maximum potential use value were applied (Gloudemans 1974). In a limited number of situations, notably in isolated, rural settings, the *ad valorem* value might be the agricultural value of a piece of property. More often, however, the value of land for agricultural use (the value of the land based on its present or potential agricultural earnings) does not approach the urban-related market value. For example, farmland with an agricultural value of $75 per acre was carried on the property tax rolls at $2,000 per acre by Montana tax assessors (Conroy 1978:13). As a consequence, Montana farmers were paying taxes reflecting the expected income (capitalization of value) for higher value but non-aplicable land uses. Not surprisingly, agriculturalists in this situation end up paying higher effective tax rates than their non-farming neighbors (Gloudemans 1974). Faced with a heavy tax burden caused by location relative to more intensive land uses, farmers have been forced to make land use changes because their businesses, no matter how efficiently operated, could not afford to operate.

Use-Value Property Taxation. To overcome the problems created by *ad valorem* procedures, most provincial and state governments have instituted some form of differential or use-value taxation. The use-value approach assesses property in its current use rather than its market or maximum potential use. The taxable value is derived from the capitalization of expected value either from rent or owner-operators' net income. Thus, agricultural land is assessed based on its value for agricultural production, reducing financial pressures.

An equally important aspect of a use-value taxation program is its political acceptability. For politicians this is an extremely attractive policy with low decision-making costs. So long as the shift in tax burden produces only slight additional costs to individual tax payers, and the benefits to recipients are clearly recognizable, political liabilities are negligible. Moreover, a use-value taxation strategy does not offend political conservatives concerned with government infringement on private property rights. Consequently, it enjoys the support from conservative interest groups opposed to traditional land development regulations, such as zoning and land use planning (Esseks 1978).

Within the broad framework of differential property tax assessment three categories of programs can be identified, including preferential assessment, deferred taxation, and restrictive agreements. While the exact provisions may vary slightly, preferential assessment provides simply that agricultural land be assessed based on its use. Currently, 17 states and 2 provinces have implemented this type of program (Table 8). The major criticism surrounding this type of use-value assessment is that it lacks a mechanism to restrain land owners from abusing the legislative intent by enjoying reduced taxes until such time as they choose without penalty to convert lands to "higher uses."

In an attempt to mitigate objectionable aspects, 25 states and 3 provinces have adopted the second category of differential assessment, deferred taxation (Table 8). The basic difference between the deferred and preferential approaches is a tax recapture clause. In implementing deferred taxation, the local assessor determines two values for each parcel of agricultural land: (1) its value in agricultural use, which serves as a basis for current taxation and (2) its value without the deferred program. If the land is subsequently converted to non-agricultural use, the taxes that would have been due without deferral are collected. Individual state programs vary in the number of years for

TABLE 8 STATE AND PROVINCIAL TAX POLICIES DESIGNED TO PROTECT AGRICULTURAL LAND

Preferential Property Taxes	Deferred Property Taxes	Restrictive Agreements	Tax Credits	Death Taxes
Arizona	Alabama	California	Michigan	Alabama
Arkansas	Alaska	Hawaii	Wisconsin	Alaska
Colorado	Connecticut	New Hampshire		Arizona
Delaware	Illinois	Pennsylvania	Ontario	Arkansas
Florida	Kentucky	Washington	Quebec	California
Idaho	Kansas			Colorado
Indiana	Maine	Ontario		Connecticut
Iowa	Maryland			Delaware
Louisiana	Massachusetts			Florida
Mississippi	Minnesota			Georgia
Missouri	Montana			Illinois
New Mexico	Nebraska			Kansas
North Dakota	Nevada			Kentucky
Oklahoma	New Jersey			Maryland
South Dakota	New York			Michigan
West Virginia	North Carolina			Minnesota
Wyoming	Ohio			Mississippi
	Oregon			Missouri
Alberta	Rhode Island			Montana
British Columbia	South Carolina			New Mexico
	Tennessee			New York
	Texas			North Dakota
	Utah			Oregon
	Vermont			Tennessee
	Virginia			Utah
				Vermont
	New Brunswick			Virginia
	Nova Scotia			Washington
	Prince Edward Island			

Sources: Coughlin *et al.* 1981; Davies and Belden 1979; McCuaig and Vincent 1980.

which deferred taxes are collected, and in some cases interest penalties are charged on the taxes.

The third approach to differential tax assessment seeks to integrate the mechanism into a broader planning framework. In a restrictive agreement program, the landowner agrees to restrict the use of his land for a specified period of time. In turn, the unit of government agrees to specified tax concessions. Typically, the land use is fixed for ten years, and either party, the landowner or government, must be given several years' notice if he intends to withdraw from the agreement. California and New Hampshire have stringent restrictive agreement programs, while Hawaii, Pennsylvania, and Washington have weaker, less penalizing programs. Among Canadian provinces, only Ontario requires farmers to commit their land to agricultural use for a specific time period, 10 years.

Tax Credit Programs. One of the major concerns surrounding all types of tax assessment is the potential impact on local government revenues. Many municipalities, counties, and public agencies rely heavily on property taxes to provide budgetary requirements. In areas where large amounts of agricultural property are assessed at lower use value, the tax burden must be shifted to other classes of taxable property or simply foregone. The most popular approach for reducing farm taxes while minimizing deleterious impacts on local government is tax credit programs for agricultural land.

First developed in Ontario (1970) and Wisconsin (1977), the concept was adopted by Michigan and Quebec. Under a tax credit program, land owners receive a reduction in property taxes, but the reduction is applied as a credit against state income taxes (U.S.) or paid as a rebate on municipal taxes (Canada). In both countries, eligibility standards have been developed to insure that only bona fide farmers participate. In the U.S. the magnitude of the benefits are tied to farmer income through a "circuit breaker" clause in the law, with farmers excused from additional property taxes once the tax bill exceeds a given percentage of their income. Where the tax credit exceeds the tax liability, a negative income tax payment is paid by the state.

The level of individual benefits in tax credit programs is generally moderate. For example, in 1978 rebate checks issued under Ontario's farm tax reduction program averaged C$308.30 (McCuaig and Vincent 1980:11), while the average tax credit in 1980 received under Wisconsin's program was US$1,413 (Coughlin *et al.* 1981:219). These costs are shared by tax payers throughout the province or state rather than borne by local government.

Inheritance and Estate Taxes. With the Tax Reform Act of 1976 the U.S. Congress made significant changes in the federal estate law permitting a number of tax reductions for eligible farm estates. Specifically, Section 2032A allowed the estimation of land value based on use value criteria for family farms left to heirs of the deceased (Coughlin *et al.* 1981:65). Moreover, Section 6166 permitted deferral of the payment of estate taxes for five years after death and then payment in installments over the next 10 years (Coughlin *et al.* 1981). The intent of these changes was to reduce high estate taxes which may force the sale of family farms because heirs do not have liquid assets to pay death taxes.

Following the federal lead, a majority of states modified their estate taxes to reduce the burden on farmland owners. Considered alone, the inheritance and estate tax reductions probably have no major effect on retarding farmland alienation. They do provide valuable support in the larger context.

Effectiveness of Tax Policies

Tax incentives have been politically attractive, as measured by the large success of these programs in both Canada and the U.S., but their efficacy in protecting agricultural acreage and maintaining farm viability has been questioned. Increasing research at a variety of scales throughout North America has shown that tax incentives alone do not prevent agricultural land from shifting into higher value uses. At best they postpone such changes.

These conclusions are supported by dozens of planning reports and empirical studies (Barlowe and Alter 1976; Conklin 1980; Gustafson and Wallace 1975; Furuseth 1980a; Toner 1978; Vogeler 1978). More importantly, comprehensive national scale research has been completed. In the U.S., the Council on Environmental Quality funded a 1976 study, *Untaxing Open Space,* to evaluate the effectiveness of differential taxation programs. The findings of this research were that "differential assessment laws in general work well to reduce the tax burden on farmers. Acting alone, however, they are not effective in preserving current uses" (Keene *et al.* 1976:115). A parallel study examining the impact of use-value taxation in Canada by McCuaig and Vincent for Environment Canada similarly concluded that ". . . programs designed to keep land in agriculture have only marginal effects, especially in areas where potential profits from non-agricultural development far outweigh possible returns from farming even with tax aid" (McCuaig and Vincent 1980:18).

The major problems surrounding tax incentive programs are twofold. First and most importantly, the financial benefits provided to land owners are slight when compared with the market value of developable land. Urban land uses will almost always outbid agricultural uses, and tax savings are not enough to make up the difference. Keene *et al.* (1976) found that when farmers were committed to maintaining agricultural land uses but facing serious financial difficulties, the savings from use value taxes might prove valuable. However, in situations where the property owner is indifferent or actively looking for an opportunity to sell his land, tax savings do not deter him from selling.

The second problem is landowner participation. Often, eligibility criteria for enrolling in the program either exclude bona fide farmers or, conversely, enable hobby farmers and speculators to qualify for tax reductions. The latter problem affects the credibility of use-value taxation. In numerous instances, use-value assessment has been used by land speculators to hold land cheaply under the guise of agricultural use, and, when the time was right, to develop and realize windfall profits. In Dade County, Florida, for example, realtors and land speculators acquired most of the remaining agricultural land in the county and rent it to farmers. Consequently, they enjoy low holding costs while awaiting proper conditions for development (Metropolitan Dade County Planning Department 1980:20-21). Speculators have been able to do this because many differential assessment programs have qualifications easily met and lack serious penalties for changing land use after gaining the benefits of low assessments. Reaction over the past several years to the corruption of program goals has been to tighten qualifying criteria to reduce cheating. One unintended impact has been to reduce participation by legitimate farmers. In some areas, large corporate or chartered farms cannot qualify for benefits; conversely, in other areas small, economically marginal (and therefore vulnerable) farms may be excluded. In both situations program goals are not met.

The problems surrounding tax-based farmland protection programs sharply curtail their usefulness. Consequently they are the least effective means of farmland protection, according to many researchers. Increasing dissatisfaction with these policies is manifest in large-scale abandonment of such tax programs during the past seven years (Furuseth 1981).

Transfer of Development Rights

A second type of financial compensation is based on the concept of development rights and the ability to move or transfer development rights (TDR). This is one of the newest and most innovative tools in U.S. land use planning. It has not been used in Canada, and may be irrelevant to the Canadian scene. The highly regarded Canadian planner Mary Rawson (1977:68) has argued:

> *The concept of transfer of "development rights" so called, is becoming a part of the planner's kitbag: TDR being imported, like the zoning concept so many years ago, from the United States. A concept more fundamental than TDR is surely already in place here in Canada, namely Crown ownership. Owners in fee simple are tenants of the Crown, and no more than that. Basic land ownership rests not with individuals but with society as a whole. TDR moves us away from that concept.*

The TDR mechanism is based on the concept that one of the "bundle of rights" of real property ownership is the right to develop the land surface. This right may be separated from fee simple ownership and sold or transferred by the owner. Just as a property owner can legally separate and sell his mineral or air rights, he can also transfer his development privileges to someone else.

Under a TDR program development rights are freely transferable between private parties or between a land owner and a public agency, with the use of these rights limited only by normal police power regulations. The costs of development rights are established by market demand. The owners of property which has been stripped of its urban development potential by public sector action are compensated for their "loss" either by allowing them to transfer development rights to the private market or by public sector purchase of the development rights (Figure 12).

In an attempt to achieve a more successful and permanent protection of farmland, an increasing number of U.S. governments are experimenting with TDR programs. Over the past 7 years, almost 30 local and state governments have adopted programs (Coughln *et al.* 1981:11-12). The most popular approach, pioneered by Suffolk County, New York, involves the community purchase of development rights to agricultural land. Title to the land remains with farmers. Seller are free to cultivate or let the land lie fallow, but cannot sell the land for any purpose other than agriculture. This form of TDR system has been used to protect agricultural land by the states of Maryland, New Jersey, Massachusetts, and New Hampshire, and by local governments in Washington and New York. As one might expect, this approach has been used almost exclusively in rural areas juxtaposed with rapidly growing urban centers.

Among the strongest supporters of these programs are farmers. For farmland owners the program provides the promise of fair compensation for surrendering their rights to develop the land. Property taxes would be permanently based on the land's value for farming (since no other use is legal), and families would inherit the land at its

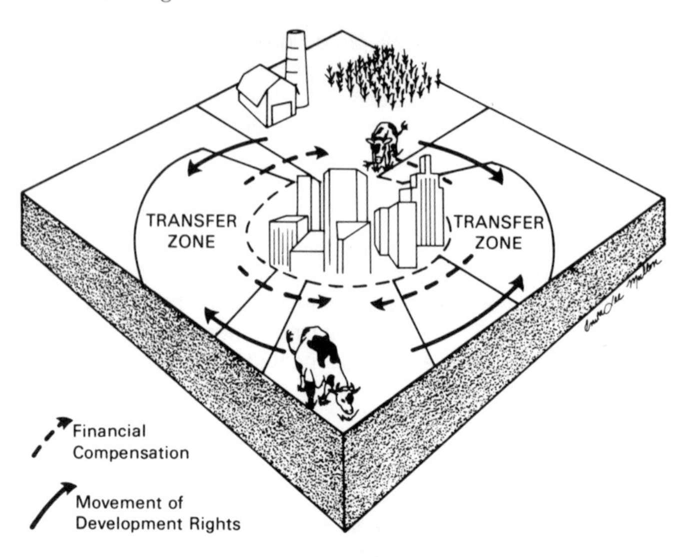

FIGURE 12 TRANSFER OF DEVELOPMENT RIGHTS.

lower farm-use value, lessening the need to sell land in order to pay death duties.

The public benefits expected to result from these private gains include supporting the survival of local agribusiness, securing supplies of fresh agricultural products for local consumers, preserving open space which will be privately maintained yet generate taxes, and curbing urban sprawl.

The major deterent to the widespread application of this strategy is cost (Esseks 1978:210). The prices paid for development rights have ranged from $3,120 per acre in Suffolk County to $311 in Burlington County, New Jersey (Coughlin *et al.* 1981:163). The price commanded reflects the amount of urban development pressure and therefore the tightening land market. In every case, the costs of purchasing sufficiently large amounts of development rights to bring about the "critical mass" of agricultural land necessary to sustain agribusiness is high (Consider the funding necessary to save only 10 percent of the agricultural land resources in your home county or regional district). Moreover, in an era of challenges to government spending, the expenditure of large amounts of money to buy development rights would not be popular.

A second approach using the TDR mechanism does not involve public purchase of development rights. Under this type of program, a given area is designated as a protection or preservation district, with only agriculture and other low intensity uses permitted. Other areas, however, within development districts can be developed at densities greater than permitted under existing zoning so as to absorb the growth which is deflected from the agricultural protection zone.

Land owners in the protection zone would be given development rights to compensate for the diminution in value of their land. Property owners in the development zone can exceed the permitted development limits if additional development rights are acquired. Thus, supply and demand for development rights would be created. In order to attain greater economic gain, land owners in the development zone would be willing to acquire the newly issued development rights. These rights are purchased, at fair market value, from farmland owners.

In spite of recent nationwide interest, this form of TDR program has been used only on a small scale in a few areas. Nationally, eight townships in Connecticut, Massachusetts, New Jersey, Pennsylvania, and two small towns in New York have experimented with the technique (Emanuel 1977; Coughlin *et al.* 1981:11). This approach has been employed with some success in other planning situations, notably the protection of environmentally sensitive wetlands in Collier County, Florida, in Puerto Rico, and in many cities to protect historic buildings (Pizor *et al.* 1979).

The major impediment to the adoption of this type of TDR program is inertia, essentially public and government hesitancy to discard long-established planning mechanisms in favor of new, untested alternatives. Landowners may be reluctant to accept development rights in lieu of the right to future development. The average citizen or elected official may not understand the system. Aggregated, these problems have caused several state governments to consider and then reject large scale TDR policies. During the past five years statewide TDR programs were proposed unsuccessfully in New Jersey, Connecticut, California, and Hawaii (*Practicing Planner* 1977).

A second concern is the creation of a balanced TDR market. Critical to program success is sufficient demand for devlopment rights. Otherwise, supply and demand will not function. The market must operate smoothly to ensure fair compensation to restrict land owners. Critics suggest that this second concern makes TDR programs unworkable.

One other very limited type of TDR program to protect agricultural lands is used in Alaska. In this state, where most of the land resources are owned either by the federal or state government, prospective farmers buy state-owned land to start or enlarge agricultural operations (Alaska 1979). These sales do not, however, include development rights. To ensure that valuble agricultural resources remain in agricultural use the government of Alaska is preempting the potential development of these lands. This use of the development rights transfer will remain uniquely Alaskan in the U.S., owing to the particular land ownership pattern of the state. However, it could be used in portions of Canada where large amounts of Crown-owned land with agricultural potential remain.

Police Power Strategies

One of the most important mechanisms for public participation in the private land use decision process is regulation. The exercise of public regulatory power falls under the police power rights of government to protect "public health, safety, comfort, morals, and welfare." In Canada the power of the province to establish police power controls is clearly articulated in the British North America Act. Generally, the provincial legislatures have delegated these powers to local, municipal, and regional governing bodies. In a similar fashion, the federal and state constitutions in the U.S. provide police power controls over private land use. These powers are then assigned to local governments through specific enabling legislation.

The specific planning tools usually identified with police power regulations include zoning, subdivision regulations, building codes, and sanitary regulations. Among these, zoning and subdivision controls have been devised to govern the locational characteristics and intensity of urban growth. The direct application of these tools to reduce agricultural losses is a recent innovation with limited distribution. In the U.S., these police power strategies are used exclusively by local governments — county, township, or municipal. In Canada they are primarily employed by local governments. However, the provincial program in Ontario is also a police power strategy.

Agricultural Zoning

Traditionally, zoning was the tool of urban planners. Municipalities favored the concept of defining individual land use districts as a way to protect property owners from externalities generated by other land uses and also to stabilize and enhance land values. In the rural areas surrounding urban concentrations land use maps usually labeled agricultural fields as "vacant" or "suitable for future development."

In many parts of North America zoning is still considered only in an urban context. Rural environments do not necessitate the stringent regulation inherent in zoning ordinances. But as cities, suburbs, and their attendant land use problems have stretched further into the countryside, the need for land use regulations has become more apparent to both politicians and land owners.

In the search for regulations to protect farmland, agricultural zoning is a popular choice. The establishment of agricultural zoning usually takes place within the context of the broader community zoning ordinance. Ideally, following the development of a general or comprehensive land use plan which guides future land use change, the community develops its zoning ordinance. This legal document specifies the land use districts in the community and establishes rules for the intensity of use in each.

In an agricultural zoning district, farming, ranching, or related uses are given highest priority. In some cases only agricultural uses are permitted (an exclusive use district), while in other situations non-farm development may be allowed if it meets specified criteria based on compatibility with agriculture (non-exclusive use district). The most important characteristic of an agricultural zoning ordinance is the extent to which it controls the intrusion of non-farm uses into an established agricultural area. If the agricultural zoning permits too much non-agricultural use, then the protective goal may be lost.

Obviously, the exclusive agricultural zoning district represents the strongest zoning program. These ordinances prohibit nonfarm dwellings or activities. The major advantage is that conflict between farm and non-farm uses is minimized. Politically, these are the most difficult type of agricultural zoning to implement. Owners of farmland and land development interests are likely to oppose police power actions which preclude future land conversion opportunities.

The non-exclusive agricultural zoning strategy is by far the most popular approach. Nonfarm land uses are permitted but agricultural uses are preferred. Usually these districts have rules designed to limit the amount and intensity of non-agricultural development. For example, large lot sizes may be required for residential development. In Dekalb County, Illinois, the minimum residential lot size in agricultural zones is 40 acres (16 ha); Napa County, California requires 100 acres (40 ha); and Weld County, Colorado, 160 acres (65 ha; Coughlin *et al.* 1981:114-116).

A more complex approach has been labeled the sliding-scale formula. Under this system, the density of non-agricultural use in the agricultural district varies inversely with the size of land ownership. For example, if a farmer owned a 20 acre tract, the sliding scale might permit him to sell or develop two half-acre residential lots, an average of one lot per 10 acres. But, if the farmer owned 100 acres, the sliding scale might permit only 4 half-acre lots, or an average of one per 25 acres. The rationale for this inverse relationship between ownership and permitted development is based on the premise that larger landholders are more important for maintaining long-term agricultural viability and, therefore, non-agricultural development on these parcels is more restricted (Toner 1978:15).

Effectiveness of Zoning

The collective experience of numerous local governments with agricultural zoning suggests that zoning is not a panacea. Rather, just as traditional urban zoning has many imperfections, so does rural zoning when it is the principal mechanism for protecting farmland.

One of the major concerns is permanence. How long before the land is rezoned? Changes in zoning districts are inherent in the zoning process. Through a number of mechanisms, including rezoning, amendments to the zoning ordinance, or zoning exceptions, an agricultural district can be subverted almost overnight.

Moreover, agricultural zoning may be effective at restricting high density urban growth, but low density growth is not controlled. If the purpose of this zoning is to maintain agricultural activities, then low density residential development is equally as bad as high density development. Large lot requirements encourage the development of "ranchettes" or hobby farms. While restricting the number of residential parcels, the net effect of large lot zoning is to remove large quantities of active agricultural land from commercial use. For example, one 50-acre hobby farm would consume as much agricultural land as a conventional residential subdivision with 100 half-acre lots. Some observers have suggested that "large lot" zoning is more destructive to the agricultural economy than is conventional higher density zoning (Conservation Foundation 1981).

Many communities have learned that zoning alone is not enough to save farmland. Agricultural zoning can be subverted by other public policies apart from the zoning process. Most attempts to control farmland alienation by relying on zoning have met with only marginal success (Toner 1978:6).

Provincial Police Power

The circumstances and actions of the Ontario's Conservative government provide a unique large-scale use of police power controls (Bray 1980). Ontario is alone among Canadian provinces and American states in its strategy. The Ontario Planning and Development Act passed in 1973 specified a process for the preparation of official regional plans (Ontario Ministry of Housing n.d.). These plans include the mapping and allocation of future land uses. Local zoning and other bylaws must be compatible with the Official Plan. The province maintains the right to review and veto these plans.

In March 1976 the Ontario Ministry of Agriculture issued a policy statement specifying the government's commitment to preserving better quality agricultural land and maintaining the economic viability of Ontario agriculture (Ontario Ministry of Agriculture and Food 1976). Subsequently, in 1977, guidelines were developed to identify valuable farmland and require local governments to alter their plans to conform with agricultural goals (Ontario Ministry of Agriculture and Food 1977). Municipalities had up to five years to revise their Official Plans.

Unfortunately, Ontario's strategy of using local planning and police power controls, guided by provincial oversight, has not worked. Generally the Official Plans and their implementing regulations have fallen victim to urban growth pressures. Critics charge that provincial agricultural policies have been distorted and destroyed by their own faulty enforcement mechanisms (Beaubien and Tabacnik 1977:56-60; Krueger 1980). Essentially, police power regulations have not provided permanent protection for farmland in the face of urban development pressures. Ontario's most valuable agricultural district, the Niagara Fruit Belt, is an area where the failure of the provincial police approach is most evident.

Comprehensive Strategies

The preceeding discussion was concerned with financial and police power mechanisms adopted to protect agricultural land. These policies generally reflect a reactive and incremental (rather than anticipatory and comprehensive) approach to problem solving. This emphasis on simple and easy solutions stems, in part, from North American land use policies in general. Blessed with an abundance of land and an almost sacred image of private property rights, government has historically promulgated land policies designed to offend the least number of people and be minimally restrictive. Consequently, land use policy has usually developed incrementally to ameliorate current problems rather than anticipate future needs. Whenever possible, reliance on the free market system to allocate land uses is preferred.

With the emergence of farmland protection as an issue, policy makers and planners sought to handle the problem in traditional fashion, with individual responses. A few governments tried incremental approaches but were forced to abandon them as land alienation continued unabated. Others realized from the outset, however, that these policies would not succeed with an issue as complex as farmland protection. They evolved comprehensive strategies which include a combination of incentives, usually tax-related, and regulatory controls to protect agricultural land resources.

Unlike financial compensation and police power strategies, the comprehensive approach applies the "carrot and stick" rationale. Property owners whose land use options are reduced by regulations to control farmland conversion are provided with an attractive set of financial benefits. If either component of the formula is missing, the policy is flawed, according to proponents of this approach. For example, if financial incentives, such as tax reductions, are applied singularly, those participating in the program are given benefits with little responsibility. They can continue to participate so long as it is financially attractive and then withdraw. The policy goal of protecting agricultural resources is eroded as land moves in and out of the program. The converse situation occurs where farmland is encumbered by regulations without financial consideration. Absence of financial aid can lead to the protection of farmland but not farmers. Faced with decreasing profit margins, North American farmers are caught in the squeeze between stable prices and higher operating costs. The value of marginal financial benefits to farm owners is becoming more important to farm survival.

Finally, a major underlying issue in public policy decision making is the question of "winners and losers" (Goodwin and Shepard 1974). Every policy action creates winners, those persons who give up the least and gain the most by an action, as well as losers, those who give up more than they gain. Comprehensive farmland protection strategies are intended to minimize the gains and losses. The public which derives benefits from protecting farmland provides financial assistance. In return agriculturalists accept regulatory controls for maintaining agricultural land uses. This insures that the farmland owners are not winners or losers in the farmland protection effort, but rather that all affected parties share the costs of the program.

Mandatory versus Voluntary Programs

There are two structural subcategories of comprehensive policies — mandatory and voluntary programs. As implied by their names, the major difference between the two is the degree of coercion. Mandatory comprehensive polices include legislative or legal mechanisms specifically compelling governments and individuals to participate. These include mandatory land use regulations which reduce the opportunity for land conversion. The restrictive aspects of these policies are tempered by financial incentives to land owners. Philosophically, governments which adopt these policies assert public primacy over private interests with respect to agricultural land issues. Once enacted, the protection of farmland enjoys equality or priority rather than secondary status with respect to other land use policies.

Viewed in terms of political feasibility, the mandatory policy is the most difficult to institute. The reason for the difficulty is clear: These programs are immediate and coercive in their effects. Not surprisingly, only a small number of governments have adopted this type of policy. These include British Columbia, Quebec, Newfoundland, and their U.S. counterparts, Hawaii and Oregon (Furuseth and Pierce 1982).

In our second program category, voluntary policies, participation is not required. Administratively, these programs are centered around enabling legislation which permits local governments to implement regulatory mechanisms and provide financial incentives to farm operators. The initiative for adopting such mechanisms, however, rests with the community or individual. In these programs, financial incentives play the important role of inducing land owners to join the program and accept the restraints on land use. The restraints assure the public that their expenditures will achieve the long-term purpose of protecting agricultural land. Because of their voluntarism, these policies are politically more feasible. The enactment of a voluntary policy can be justified to opponents by pointing out that participation is not required, while farmland protection supporters are encouraged by regulatory aspects of the policy. It may be that this strategy represents the politically ideal policy, one which offends few persons and ultimately provides political payoffs for decision makers.

This type of policy is used in provincial, state, and local governments in Wisconsin, Virginia, Minnesota, Connecticut, Maryland, Massachusetts, Illinois, New Hampshire, and Prince Edward Island (Furuseth and Pierce 1982). At least three other states — Delaware, Ohio, Pennsylvania — are currently amending their protection programs and will soon join this group (Furuseth 1981).

Integrated Provincial/State Programs

One model for comprehensive farmland protection is a broadly based, integrated program implemented at a provincial or state level. Since the late 1960's, state-wide land use planning legislation has been a significant feature in America's "Quiet Revolution in Land Use Control" (Bosselman and Callies 1971). This trend has made it possible to develop specific comprehensive policies for farm and ranchland on a large scale.

Central to this strategy is provincial or state government legislation articulating an official policy for protecting agricultural resources and establishing the mechanisms for implementation. The regulatory aspect focuses on the control of agricultural land use through agricultural zoning, complemented by financial incentives provided as tax rebates or discounts on the affected farm property.

Key characteristics of these policies are their balanced approach, combining restrictions on land use with economic benefits to farmers, and leadership by provincial and state governments. Recognizing that local government may be parochial and too accommodating to local development interests, the role of higher government is to promote regional welfare and long-range land use objectives. A strong provincial or state involvement underlies all local planning involving agricultural land. Implicitly, regional interests dominate over local interests in the area of agricultural land planning.

The earliest U. S. program of this type was enacted by Hawaii (1961), followed by Oregon (1973) and Wisconsin (1977). Hawaii has state agricultural zoning (Bosselman and Callies 1971), while Oregon and Wisconsin allow local governments to control agricultural zoning so long as state standards are met (Barrows and Yanggen 1978; Furuseth 1980b). Both Hawaii and Oregon use deferred differential taxation for agricultural land and Wisconsin employs a circuit-breaker tax formula.

In Canada, British Columbia was the innovator (1972) with strong provincial agricultural zoning and differential taxation on farm property (Pierce and Furuseth 1982). More recently, Quebec and Newfoundland have followed the B. C. example.

Agricultural Districting

Another comprehensive strategy is agricultural districting. Here, agricultural activities are protected and assisted in a voluntary special district. The concept was originally developed by the State of New York as local-option legislation (Conklin 1980:6). Districts are initiated by local landowners and must encompass at least 500 acres (202 ha), except in Maryland where the minimum threshold is 100 acres (40 ha; Nielsen 1979:440). To be designated, the proposed lands must be approved by local and state government. Districts are created for a fixed but renewable period of time ranging from 4 to 10 years.

Generally, the provisions of the program oblige the farmers within the district boundaries to maintain agricultural use of their land during the contract period; in return they receive a number of benefits. Benefits include preferential property tax assessment, protection from local government ordinances hindering normal farming operations, limits on public infrastructural investments that would promote non-farm development within the district, waiver on special tax assessments for capital investments, and state agency regulations supporting agricultural land uses (Lapping 1980:162).

The agricultural district concept is based on the proposition that if farmers are provided with incentives to join in the creation of protected districts where farming is the only activity, and if they are protected against "outside" factors which make farming undesirable or unprofitable, they will keep their land in agriculture. The formation of an organization initiated by farmers will, it is hoped, strengthen the future of agriculture in the area.

Since its inception in New York, enabling legislation to permit the formation of agricultural districts has spread to four other states, Virginia (1977), Maryland (1977), Illinois (1979), and Minnesota (1980; Lapping 1980). One other state, California, has legislation fostering the district concept. In practice, however, the law has resulted in individual farms rather than groups of farms constituting districts (Geier 1980). The Minnesota legislation affects only the Twin Cities (Minneapolis-St. Paul) metropolitan area.

The voluntary nature of the agricultural district has made it politically acceptable to farm property owners and development interests. Its low operating cost has made it acceptable to the general public. However, evidence collected in New York, where sufficient longevity permits evaluation, indicates that agricultural districting has been relatively ineffective in reducing farmland alienation. Incentives and regulations are not sufficiently strong to operate as intended in areas of major urban development pressure

(Conklin 1980:8-9). They are better suited to handle incipient urban pressures in rural environments. Nevertheless, measured against non-comprehensive strategies, agricultural districting is far more effective.

Measuring Policy Effectiveness

Every political unit in Canada and the U.S., whether local, regional, or provincial/state, faces a different set of physical and political factors critical to the enactment and ultimate success of a policy to protect agricultural land. Consequently, it is probable that no single farmland protection program will be suitable to all settings and conditions. We can, nevertheless, suggest why specific policy actions generally operate better than others. Implicit in these evaluations is the notion of policy efficacy. How does one define a successful versus a faulty land use policy?

The answer is not simple. Policy effectiveness can be measured along several different dimensions. Among the criteria employed are land use performance (the deterence of unwanted land use change), political acceptability, and longevity. While these factors may be weakly interrelated, each operates independently. Hence, a policy may be judged effective using one criteria but fail using another. This is especially the case when comparing land use effectiveness and political popularity. Unfortunately, research requirements to answer these questions are beyond the scope and funding of most planning agencies. As a result, the land use dimensions of program effectiveness are not readily available for most provinces and states. Census records are the data most often available. They provide some insight but are not oriented to measuring policy effectiveness.

The political perspective has dominated current policy evaluation (Mundie 1980). In some cases, merely the act of implementing a specific farmland protection program is interpreted as a policy success. This preoccupation with public acceptance partially reflects the difficulty of implementing an agricultural land protection program. Esseks (1978) theorized that in order for a policy to be implemented and survive it must enjoy the support of farmers and non-farm allies, usually environmentalists and advocates of slow growth. The collapse of this coalition will lead to policy failure or withdrawal. Bryant (1975) drew broader conclusions, asserting that the general public must be convinced of the need to protect farmland and the effectiveness of a particular approach. Moreover, he suggested that the public is interested in how a program will affect them and what its cost will be. In either case, public support is necessary not only for the enactment of a program but also to maintain program goals. Without continuing interest and vigilance, policies may become stagnant and "bureaucratized" or be amended and modified to reduce their impact. For these reasons, public support and program longevity are also important measures of policy effectiveness.

6

Studies in Farmland Protection

The enactment and implementation of public land use policies is never easy. Transfer of theoretical concepts and legal rules into practice invariably generates unexpected effects. The experiences of four communities which have enacted farmland protection policies illustrate impacts of these policies on the social, economic, and political environment. Our analysis draws on planning reports, published research, personal communication, and, in several cases, visits to the jurisdiction.

The first study area, the Niagara Fruit belt of Southern Ontario, represents the application of a regional, indirect police power approach to reduce farmland alienation. Maryland, the second case study, is an example of an evolving farmland protection policy. Faced with continued erosion of agricultural land, Maryland has gradually sought stronger and more complex policies. King County, Washington, demonstrates the use of financial compensation methods, notably development rights purchase, to preserve farmland in a metropolitan environment. Finally, the province of British Columbia stands out in its use of a mandatory, integrated program which employs police power controls and financial incentives to stop the alienation of farmland.

The Niagara Fruit Belt: A Soft Approach

Few areas in Canada are suitable for the production of tender fruit crops. The Okanagan Valley in British Columbia, the Kent-Essex area of southwestern Ontario, and the Niagara Fruit Belt on the south shore of Lake Ontario represent environments with the greatest potential for fruit production. A closer examination of each area reveals that the Niagarga Fruit Belt possesses the optimum combination of soil and climatic conditions in Canada for fruits, and it is this advantage which has contributed to the region's importance as the country's premier producer of tender fruits — peaches, grapes, cherries, pears, plums, and small fruits. Because of its superior soils and low risk of frost damage, Krueger (1977a) has argued that it is second only to California in terms of natural environment for tender fruit production.

As in the case of California, superior natural conditions do not guarantee successful and continuing farm production. The Fruit Belt's flat, well-drained land; its use as a transportation corridor; its proximity to the metropolitan regions of Toronto, Hamilton, and St. Catharines; and decentralized growth of both industrial and residential uses of the land have combined to threaten the very existence of its agricultural base (Figure 13).

Over the past thirty years urbanization has brought significant changes. Agricultural development peaked in 1951 with some 32,600 acres (13,170 ha) planted in tree crops and 20,400 acres (8,241 ha) devoted to vineyards. By 1980 less than half of the original tree crop acreage was still in production, but grape acreage had expanded by approximately 6,000 acres (2,424 ha). Total agricultural acreage declined by 20 percent during this 30 year period — the majority of which was tender fruit soils. Although some losses have been offset by intensified use of remaining farmland, prospects for

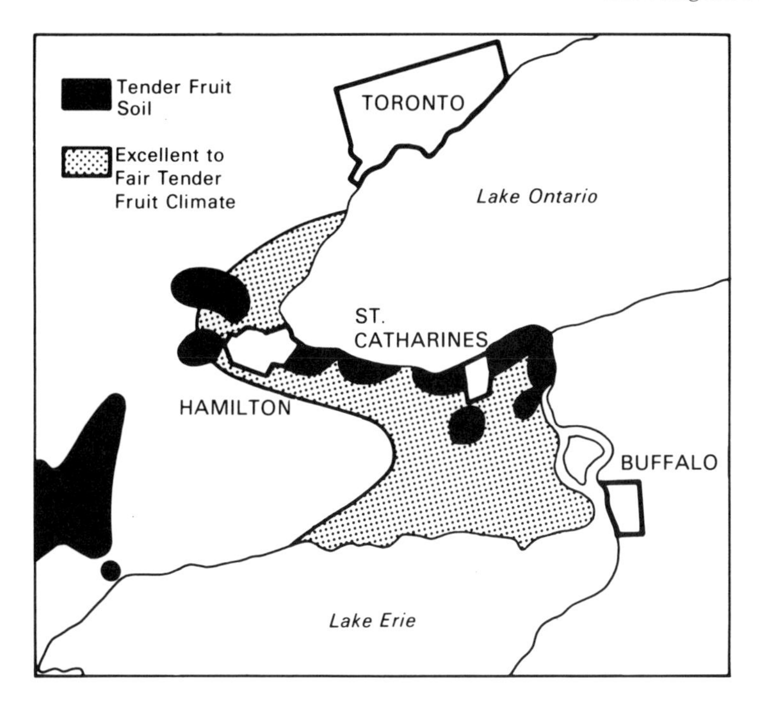

FIGURE 13 NIAGARA FRUIT BELT (Map courtesy of Dr. Ralph R. Krueger, University of Waterloo. Map reproduced from *Canada's Special Resource Lands;* Simpson-Lewis 1979).

increased fruit production are slight since practically all of the tender fruit acreage has been planted (Krueger 1978).

The impact of urbanization on the processing industry have been even more severe. Today the Niagara Fruit Belt is served by only one major tree fruit processing plant. According to Krueger (1978:190), "if it should cease operations because of lack of adequate fruit supply or because of competition from imports, then the complete collapse of the tree fruit-growing industry would come quickly."

Policy Response to the Urbanization of Farmland

During the last twenty years Ontario has relied upon relatively conventional methods for the regulation of rural land use. At best these methods are only indirectly aimed at preserving agricultural land. Under the province's Planning Act, municipalities are required to provide Official Plans which may contain designated agricultural areas. The integrity of the agricultural areas is to be enforced through zoning bylaws and severance control. The fate of agriculture within these areas is therefore very much dependent upon the nature of the severance control and the ease with which the official plan can be amended.

Specific concern for farmland preservation was voiced in the *Design for Development: The Toronto Centred Region* (Ontario 1970), in which one of twelve goals was to minimize non-agricultural use of productive agricultural land. Four years later the Central Ontario Lakeshore Urban Complex Task Force (Ontario 1974) outlined a number of imperatives and priorities ranging from maximum maintenance of producing prime agricultural land to establishing long-term agricultural use priority. However it was not until the Ontario Ministry of Agriculture and Food released its Green Paper on *Food Land Guidelines* (1977) that specific criteria were provided to assist local governments in designating agricultural land within their official plans. Some of the policy concerns for high priority agricultural land are as follows:

> 1) *The lands within the agricultural designation or to be available for agriculture on a long term basis . . .*
> 2) *The types of activity or land use that will be permitted within agricultural designation should be indicated . . .*
> 3) *Reference to the Code of Practice and the Minimum Distance Separation Formula should be incorporated into the Official Plan . . .*
> 4) *Policies should be included with regard to utility and communications facilities to ensure that their impact on agriculture is minimized . . .*
> 5) *Policies should be outlined for severances within the area designated agricultural . . .*
> 6) *Urban development such as expansion of town or rural residential growth should be in separate designation from high priority agriculture.*

Despite these guidelines, the provincial government continues to allow most land use decisions regarding agricultural land to be made at the local level, where the market place is still the most effective allocator of land. Therefore, economic development will inevitably consume good quality agricultural land.

The Niagara Region Growth Boundary Controversy

Since 1950 the decline in agricultural acreage has been due less to the growth of population and urban economy of the region and more to the distribution and location of development. Proliferation of rural, non-farm residences, low-density suburban development, and strip-commercial development represented the new form of growth which matched the increased mobility of urban society. The lack of coordination in land use planning within and among municipalities contributed to these new development trends. Although researchers during the mid-1950s had recommended regional transportation, urban services, and urban growth planning for the Niagara peninsula (Irving 1957), only in 1970 was regional government and planning instituted in Niagara.

To be in compliance with the provincial Planning Act, the Regional Municipality of Niagara was directed to produce a regional plan within three years and to submit it for approval to the Ontario government. Agricultural preservationists were hopeful that finally a mechanism had been established to save the fruit lands. Their optimism was short-lived. The 1974 plan submitted by the Regional Municipality contained a strong endorsement of agricultural development, acknowledging the importance of the agricultural sector to the regional economy and the need to protect it; however, the boundaries proposed for future urban development encompassed vast amounts of prime agricultural land. The proposed urban area boundaries were far in excess of requirements for the 1990s. Gayler (1982) argued that the projections for urban land use requirements were based upon simple extrapolation of trends in the 1950s and 1960s and did not take into account the slowing of growth in either population or the regional economy during the 1970s. He noted that "while the population projected for 1991 was at best 510,000, the proposed urban area could accommodate 640,000 people" (Gayler 1982:167). Clearly, local governments were not prepared to redirect growth nor were they willing to make the necessary adjustments to their own official plans to conform with the spirit of the regional plan. The ensuing delay in plan approval stemmed from the unwillingness of municipal and regional politicians to face the reduced urban land requirements for the region into the 1990s (Gayler 1982:169).

Following a provincial election in 1975 in which the designated urban area boundaries for the Niagara Region had become an important political issue, the provincial government returned the regional plan for revision. The regional planning staff recommended that the urban area boundaries be reduced by 2,800 to 3,200 ha (6916 to 7904 acres). This reduction would still allow a very generous area, especially in view of the expected rise in development densities. The Regional Council rejected this recommendation, proposing instead a reduction of approximately 202 ha (500 acres). Coincident with these events the Ministry of Agriculture and Food published its 1977 Green Paper, *Food Land Guidelines.* Although the 1977 guidelines were not adopted as policy until 1978, their existence obligated the government to reject the revised proposal and to propose a roll back of 1214 ha (3000 acres). However, as Krueger (1981a) noted, this compromise solution pleased neither the expansionists nor preservationists. The Minister of Housing then referred the boundary dispute for adjudication to the Ontario Municipal Board (OMB).

Starting in late 1978 and for over two years, the OMB listened to testimony from two divergent groups, the pro-expansionists (who included the local and regional governments as well as owners of seventy-five separate parcels) and the conservationists (led by the Preservtion of Agricultural Lands Society, PALS). From the outset, the pro-expanionists clearly had the advantage in terms of financial support, access to legal council, and strength of numbers. Generally, local municipalities and developers wished to retain the earlier urban area boundary on the basis that the land was too fragmented and expensive to farm, local government service plans had been made, and, finally, that the more land set aside for expansion the greater the probability of retaining the planned urban area boundaries (Krueger 1981b:9). However, PALS contested each parcel of land on the basis of government policy as outlined in *Food Land Guidelines.* Krueger (1981b) noted that questions such as "What is the proof that the land is needed for this particular urban use?" and "What considerations have been given to an alternative location on less valuable agricultural land?" put the pro-expansionists on the defensive. For the first time pro-expansionists had to prove 'why their development could not go elsewhere' (Krueger 1981b:11).

Two rounds of OMB hearings resulted in the decision to save an additional 1600-2000 ha (4000-5000 acres) of land from urban designation. Notwithstanding this victory, the OMB designated more land than was actually required for urban development. It did this to ensure that boundaries adjacent to fruit land would be permanent.

The OMB Niagara urban area boundary decision is important for other reasons. Perhaps most important is that the OMB advocated a very restrictive consent policy (severancing) outside the urban boundaries in tender fruit and grape lands, and a "slightly less restrictive" approach for high quality agricultural areas (Jackson 1982:173). In terms of the permanency of the existing boundary, the OMB argued that changes should not be made unless two conditions existed: (1) that the proposed use was essential, and (2) that lower capability land was not available. Fragmented or abandoned land were not considered sufficient reason for develoment. In this case the OMB recognized the priority of soil capability or potential as opposed to existing levels of production. Finally, Jackson (1982:175), in interpreting the OMB decision, argued that "the necessity of new development on high priority agricultural land be justified, not in relation to each municipality but on a regional basis." Hence urban development whenever possible should be directed toward areas of lower agricultural capability. This redirection can only be achieved through the mutual cooperation of the municipalities involved which is no more than "sound regional planning principles of land conservation" (Jackson 1982:173).

Conclusion

The events leading up to the OMB hearings, the hearings themselves, and the decision of the OMB regarding the tradeoffs between urban expansion and the protection of prime agricultural land have provided important insights into the 'place' of agricultural land in a largely urban society and the prospects for conservation of land in other areas of Ontario. Without the organized and consistent opposition of PALS to the urbanization of fruitlands on the Niagara Peninsula there might have been no OMB hearings, little or no public debate regarding a vital land resource, and certainly a much greater retrenchment of the agricultural base. Throughout the hearing process the ability of PALS itself to participate was often in jeopardy owing to financial constraints. If a well organized and widely supported advocacy group is required to force social action in a highly valued and unique agricultural area, what will be required in regions of significant but lesser importance for agriculture?

The publication and acceptance as policy of the *Food Land Guidelines* (Ontario 1977) is clearly an important step toward achieving a 'rational' balance between competing uses of the land. Yet its ambiguity, its reliance on the market mechanism in allocating land, and the basic decentralized land-use designation process ensure very uneven protection of farmland across Ontario. Clearly, in a province which has benefited so much from its agricultural industry, a more comprehensive, consistent, and just policy toward farmland protection is needed.

Maryland: An Early Innovator Forced to Change

In 1956 the Maryland General Assembly enacted legislation reflecting increasing concern over the state's agricultural future. It required the tax assessment of farmland on the basis of agricultural use alone. The strength of legislative resolve surrounding this issue is reflected in the fact that this law was passed over the veto of the governor (Nielsen 1979). By this action, Maryland became the first American state to take direct steps to reduce farmland conversion. Subsequently, the Maryland strategy of differential property taxation has emerged as the most widely used approach for handling this problem.

Historical Background

In the years that followed enactment, proponents of agricultural preservation in Maryland continued to "fine tune" the original legislation to increase its effectiveness. Despite these attempts, farm interests and conservationists began to recognize as early as 1967 that more comprehensive and stringent state policies were needed to guard farmland (Nielsen 1979). Much of this concern arose out of Maryland's unique urban and agricultural setting. Located on the southern edge of the United States' northeastern urban region, much of Maryland's best agricultural land is within the commuting fields of large metropolitan areas, most notably Baltimore and Washington. The impacts of the suburban land market extend throughout much of the state.

Between 1950 and 1970 Maryland's population increased over 50 percent to almost 4 million. This increase was accompanied by a 40 percent or 1,600,000 acre (646,000 ha) decline in the amount of agricultural land. The rate of loss was 1 acre for each additional state resident (Maryland Agricultural Land Preservation Foundation 1980). More recent analyses suggest Maryland's agricultural economy will continue to face increasing stress from urban land use pressures over the next 20 years. A report by the Maryland Department of Agriculture predicts, for example, a potential yearly loss of 50,000 acres (20,200 ha) of farmland through the year 2000 (Committee for the Preservation of Agricultural Land 1973).

Maryland farms supply an estimated 55% of the state's agricultural needs and provide employment for 38,000 people (Maryland Cooperative Extension Service 1978). The Maryland Cooperative Extension Service study estimated net farm income in 1977 as $101.7 million. Agriculture remains a viable component of the economy of this small state, but urban sprawl seriously threatens future agricultural values.

The Beginning of Change

As pressure mounted on the state government to take stronger action to reduce farmland alienation, a bill passed in the late 1960s created a special commission to develop a long-term plan for the preservation of prime agricultural land. In the end, the commission devoted most of its energies to recommending changes in the differential assessment law and failed to propose any specific solutions for dealing with land conversion. The commission's criticisms did, however, provide the impetus for change (Committee for the Preservation of Agricultural Land 1973).

In the six years that followed, a number of individual proposals designed to curb farmland loss were introduced in the General Assembly, but none were enacted. According to observers, a major obstacle in almost every case was local government objections to potentially greater state control over local land use (Nielsen 1979; Schiff 1979). Finally, in 1974 Senate Bill 715 was enacted. This bill created the Maryland Agricultural Land Preservation Foundation to coordinate a statewide, voluntary agricultural protection program (Maryland Agricultural Code 1974). During its early history, the Foundation was ineffectual, due to an absence of supporting legislation and funding difficulties (Nielsen 1979; Schiff 1979). These problems were corrected in 1977 with the enactment of additional legislation to complement and strengthen the Foundation. With the 1977 legislation Maryland finally abandoned its reliance on differential property taxes and embarked on an integrated strategy. This shift repesented a significant change in state policy toward agricultural land use (Nielsen 1979:438).

Voluntarism and Local Control

Maryland's present farmland protection is comprised of two components — development rights easements and agricultural districts. The former, a transfer of development rights strategy, was designed to improve the financial position of farm operators while protecting the long-term future of high-quality agricultural land. In contrast, the district concept has the short-term objective of protecting existing farming operations from externalities generated by nearby urban neighbors.

The final approval of a state agricultural land preservation program followed a number of political compromises. The program is voluntary, guarantees compensation for the affected farmland, and, thus, is satisfactory to farmers. It is acceptable to local governments as well because it requires local government involvement and approval. Consequently, county officials continue to retain control over local planning and land use decisions. With the support of these two groups passage of the compromise program was assured.

At the center of Maryland's program are the Maryland Agricultural Land Preservation Foundation and individual county Agricultural Preservation Advisory Boards. The Foundation is administered by a nine-member board of trustees appointed by the governor (Maryland Cooperative Extension Service 1977). Five of the nine must be either active or retired farmers and must represent different geographical areas of the state. Board members serve 4-year terms. The powers and responsibilities of the Foundation include purchasing development rights easements and developing rules governing the establishment of, and approving applications for, agricultural preservation districts.

The legislation creating the Foundation requires that counties organize local Agricultural Preservation Advisory Boards. Each board consists of five members, three of whom are commercial farmers. The powers of the Advisory Board parallel those of the larger Foundation on a smaller scale. They advise county decision makers in establishing agricultural districts and approving the purchase of development easements; advise the foundation concerning county priorities and needs for agricultural preservation; monitor the status of agricultural districts and lands under easement; and, finally, serve an educational function promoting the preservation of farmland within their counties (Maryland Cooperative Extension Service 1977). Essentially, the county advisory bodies carry the weight of implementation at the local level. Their familiarity with local agricultural conditions and rapport with county politicians are essential to overcoming local skepticism and criticism. An active and effective advisory board can 'sell' farmland protection in a county.

Development Rights Easements

Statutes approved in 1977 and 1979 authorize and provide funding for the Foundation to purchase development rights easements on farmland in agricultural districts (Nielsen 1979). The goal of these purchases is to ensure a source of future food for residents, to control urban sprawl, and to provide for the protection of agricultural land as open space. Although the enabling legislation was not specific, it was the intent of the General Assembly that the easements sold to the Foundation would be held in perpetuity (Nielsen 1979:447). Maryland's development easement program therefore qualifies as a transfer of development rights program.

In future cases where agriculture is no longer economically feasible, the Maryland program does permit termination of the easement agreement. The minimum period of the contract is, however, 25 years. If termination is approved by the Foundation and the local government, the land owner will be permitted to re-purchase the easement (development rights) by paying the difference between the fair market and the agricultural value at the time of resale.

Farmland owners wishing to sell the development rights of their farmland make formal application to the Foundation. Each application contains the seller's asking price as well as a description of the current use of the land and its soil characteristics. Upon receipt, the Foundation has appraisals made and submits the application to the county government for its approval. In deciding whether it should approve the proposed application, the county consults with the local Agricultural Preservation Advisory Board. Public hearings may be requested but are not required. In arriving at their decision, the county government considers factors such as local agricultural land preservation priorities, the effect on surrounding land use, and local zoning or other related regulations.

Even if approved by the county, the application can still be rejected by the Foundation. The major criteria for acceptance are the asking price and the ranking of a proposal. The price that can be paid for an easement is either the asking price or the difference between the fair market value of the land and the agricultural value as determined by appraisal, whichever is lower. This formula is mandated by the original legislation.

Because funds to purchase easements are limited, the Foundation ranks purchase priorites, with an over-all goal of maximizing the amount of agricultural land preserved at the lowest cost to the public. Prior to extending purchase agreements, all proposals during each fiscal year are ranked using the following formula (Maryland Cooperative Extension Service 1977:9):

$$\text{Ratio} = \frac{\text{easement value} - \text{asking price}}{\text{easement value}}$$

where easement value = fair market value - agricultural value.

As a product of the computation, the "lowest bidder" has the highest ratio. This formula permits a comparison of the various easement offers, including different types of agricultural land from various parts of the state. Offers to buy easements are first made to the landowners whose ratios are the largest.

With the sale of a development rights easement, a deed containing covenants relating to the use restrictions is recorded in the land records of the county. The land owner agrees to keep his farmland in agricultural use and is forbidden to use it for residential, commercial, or industrial development. Under the agreement, the Foundation is permitted to make inspections, upon notice, to ensure compliance with terms of the sale. The landowner remains free to sell or convey his property, but the terms of the easement remain in effect for a minimum of 25 years.

Agricultural Districts

The second mechanism of Maryland's farmland preservation program is farmer-initiated agricultural districts. The procedure for establishing a district is carefully outlined in the authorizing statute, passed in 1978 (Maryland Agricultural Code 1978). One or more land owners must file a petition with the county government. To qualify, the proposed agricultural district must consist of land which is either currently used primarily for agriculture or is open space capable of agricultural use. The size of the proposed district must be larger than 100 acres (40 ha), unless unique micro-climatic, topographic, or hydrologic variables permit smaller areas to be economically viable for agriculture (Nielsen 1979:440). A majority of the land must consist of Soil Capability Classes I, II, III, or U.S.D.A. Woodland Classes 1 and 2. Exceptions for lower soil capabilities are allowed in cases where specialized or unique agriculture exists. Finally, the proposed agricultural district cannot be within the boundaries of any water and sewage service district or areas already programmed for urban growth by a county or city. Petitions meeting these requirements are referred to the county Agricultural Preservation Advisory Board and to the local planning and zoning body for a report as to the advisability of establishing the district.

In its review the Advisory Board is primarily concerned with the impact that the proposed district will have on continuing agricultural production in the area. From the perspective of the planning and zoning body, the major concern is whether the proposed district will be compatible with state and local plans and programs. If either group recommends approval, a public hearing on the proposed agricultural district is held by the county government. Following the hearing, the county government recommends approval or disapproval to the Foundation. A negative decision by the county kills the petition. To date, all endorsements by county governments have been approved by the Foundation.

If the petition is approved by the Foundation, the county establishes the district through passage of a special ordinance identifying the agricultural district and promoting normal agricultural activities within the district. The ordinance, for example, must permit all types of farming, day and night operation of agricultural equipment, and all "normal and acceptable" farming practices. County governments are also encouraged to relax local building and zoning requirements on farm structures or activities that interfere with agricultural production. These protective provisions, unavailable to agriculturalists outside of agricultural districts, are among the main advantages of participation. Another advantage is that farmland must be within an agricultural district in order for its owner to sell a development easement to the Foundation.

In return for these benefits, land owners in designated agricultural districts contractually agree to maintain their land in agriculture for at least five years. The district may continue indefinitely; however, it can be terminated after the minimum enrollment period (5 years) or in the event of severe economic hardship (Maryland Cooperative Extension Service 1977:11). The notice of intent to terminate the District requires a one year notice. Additions to any existing agricultural district can be made at any time with no minimum size requirement.

Program Effectiveness

It is still too early to predict how well Maryland's most recent farmland protection legislation will perform. Nevertheless, some preliminary observations and analyses can be made. From the outset, adequate financial support has been a concern. State funding for the Agricultural Land Preservation Foundation and the purchase of development rights easements did not immediately follow the passage of the enabling legislation (Nielsen 1979:444). As a consequence, the program has been fully operational only during the past three fiscal years (Musselman 1981). Starting slowly, public acceptance and farmer participation has increased markedly. During the first year only 47 agricultural districts were formed, encompassing 12,053 acres (4,869 ha), not an encouraging beginning (Maryland Agricultural Land Preservation Foundation 1980:23). In fiscal year 1981, however, the number of districts jumped to 175 and included approximately 45,000 acres (18,180 ha; Musselman 1981).

Agricultural districts have been formed in 17 out of Maryland's 23 counties (Figure 14). According to Alan Musselman, Executive Director of the Preservation Foundation (1981), farmers were becoming increasingly aware of the protection offered by the

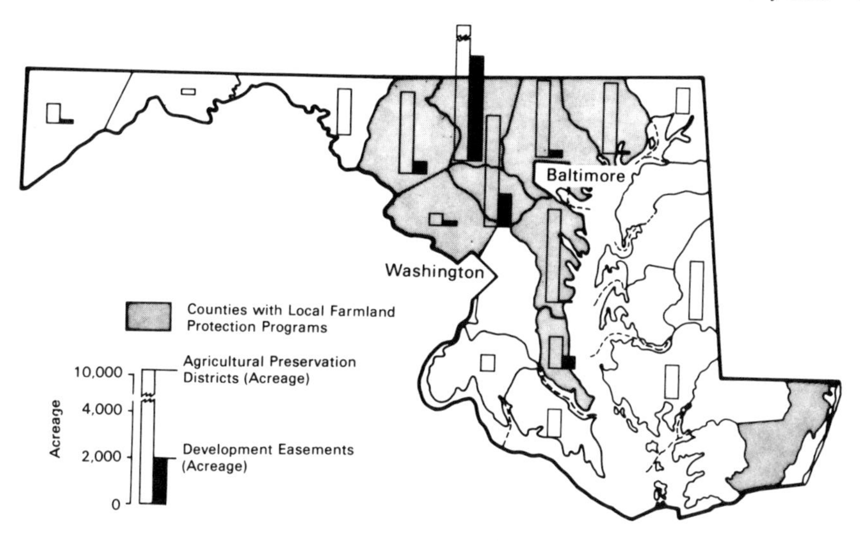

FIGURE 14 MARYLAND FARMLAND PROTECTION ACTIVITY.

agricultural districts against site selection for hazardous waste and solid waste landfills and from state and local capital improvement projects. Furthermore, he suggested that in many areas, agricultural districts are receiving "automatic protection" against government policies and actions which under normal operating procedures would destroy agricultural land values.

Similarly, the number of development easements has also increased dramatically. During fiscal year 1980, only 2,240 acres (905 ha) were included in the easement program (Maryland Agricultural Land Preservation Foundation 1980:23). By 1981, participation increased to 5,544 acres (2,240 ha; Maryland Agricultural Land Preservation Foundation 1981:1). Another 13,716 acres (5,541 ha) were in the process of easement acquisition (Musselman 1981). Obviously, this is still a small component of Maryland's total agricultural resource base, but it does represent an encouraging beginning. And, perhaps more importantly, these easement lands are concentrated in urban fringe environments where urban pressures are most intense.

Beyond its direct impact, Maryland's active state farmland protection program has encouraged county and municipal action. Currently, nine counties have implemented complementary agricultural land protection efforts, including exclusive agricultural zoning, transferrable development rights systems, and local easement programs (Maryland Agricultural Land Preservation Foundation 1981:1). Aggregated, the impact of parallel state and local policies greatly strengthens the efforts for farmland protection.

Despite early progress several serious concerns surrounding program design have been raised. Perhaps the most immediate is funding. State financing, critical for purchasing development easements, has never exceeded $6 million annually (Nielsen 1979:444-445). Monies available for fiscal year 1982 were approximately $4.3 million, with roughly $3 million in local matching funds (Musselman 1981). Future funding was not expected to increase. Yet, state officials estimated that full implementation would require $10 to $15 million a year (Nielsen 1979:444). With rising land values, the Land Preservation Foundation is currently paying an average of $872 per acre for development easements (Maryland Agricultural Land Preservation Foundation 1981:2). Given present, as well as expected, funding levels, it is unrealistic to anticipate large amounts of farmland being included under easement protection.

A second shortcoming stems from the voluntary nature of the program. As presently organized the program lacks major incentives to induce participation among farmers in an agricultural district or the easement program. As a result, one might expect decreased program involvement as the economic rewards of non-agricultural land use increase in magnitude. This may account for the already low participation rates in Montgomery and Prince Georges Counties, suburban counties north and east of Washington (Schiff 1979).

There is modest optimism that Maryland's program deficiencies will not dilute total effectiveness. Hopefully, the willingness of decision makers to modify and strengthen past legislation in face of policy failure will continue, and the weaknesses in the current program can be removed. The process of policy "trial and error" may be frustrating, but it is an integral part of the democratic system.

King County, Washington: A Regional Development Rights Purchase Program

King County, Washington, is an urban county at the center of the Seattle metropolitan region, an area with a population of over 1.2 million. Its recent history illustrates how local government can stabilize an agricultural base if the voters and political leaders are persistent and willing to pay the financial price. This county has chosen to protect its agricultural land use options through a locally funded program to purchase development rights.

Background

Agriculture in the broad valleys surrounding Seattle has followed the pattern observed throughout North America. Since the settling of the Puget Sound, large scale commercial agriculture has been a part of the King County Landscape. Over the past 25 years agricultural activities have given way to idle land and urban uses as land values have increased. Between 1959 and 1974 the number of county farms dropped from nearly 3,000 to less than 1,400. Concurrently, farm land decreased from over 114,000 acres (46,000 ha) to less than 55,000 acres (22,200 ha). Employment opportunities in food processing and other agricultural industries suffered similar declines. Despite the losses, King County farms still generated more than $40 million of primary economic income annually and provided an estimated 8,000 jobs (King County 1977:1).

Policy Evolution

Public recognition of the importance of King County's agricultural land and consideration of policies to protect the resource date back to 1964 (King County, Office of Agriculture 1977). In that year, the county land use plan established specific areas as "agricultural" zones and recommended their retention in this use. The local initiative received support from the state government in 1970 with enactment of a statewide differential property taxation program, the Open Space Tax Act (Dunford 1981:19).

In spite of these early efforts, the loss of agricultural land continued during the early 1970s. Consequently, in December 1975, the county adopted a one-year moratorium on further development of farmland so that the problem could be studied and a comprehensive program formulated (King County, Office of Agriculture 1977). During the moratorium, plans were made which set in motion the current development rights purchase program.

Adoption of County Ordinance 3064 in 1977 marked the beginning of the program. (Dunford 1981). This law identified areas of King County where agriculture was still economically viable, and within these pinpointed the best agricultural land. The ordinance designated two special planning categories, "King County Agricultural Districts" and "Agricultural Lands of County Significance" (King County Council 1977:4-6). The designated Agricultural Districts represented eight areas of the country where agricultural activities were concentrated. Within the Agricultural Districts, the ordinance identified and labelled 32,500 acres (13,130 ha) of high-quality agricultural land as Agricultural Lands of County Significance (King County Council 1977:7). These areas shared characteristics which made them both physically and economically well suited for agriculture. Among the criteria for designation were prime agricultural soils, current non-urban land uses, an absence of urban utilities, and location outside existing municipal boundaries. Parcels also had to be over 20 acres (8 ha) in size. The purpose was to separate high-quality farmland with a future in agricultural use from those lands already impacted by or committed to urban development.

Agricultural land included in the Lands of Significance category was subject to prohibitions and guidelines designed to protect its agricultural value. Specific actions included a moratorium against water and sewer extensions into these areas, large minimum acreage requirements for land subdivision, a restriction on municipal annexation of designated lands, and, most importantly, a freeze on rezoning requests. Taken together, these actions were designed to stop temporarily the loss of King County's most valuable agricultural land.

The provisions of Ordinance 3064 were designed to be only temporary (King County Council 1977:11). The ordinance was scheduled to expire at the end of 18 months. During the interim period, it called for the county to evolve and implement a permanent solution to the problem.

During the next year and a half, eight citizen advisory committees composed of farmers, agricultural industry representatives, real estate and building interests, environmentalists, and civic leaders worked with a newly created King County Office of Agriculture to evaluate alternative land management methods and prepare a specific proposal for county government consideration. Following a review and analysis of protection strategies applicable to King County, the committees proposed that a voluntary development rights purchase program be adopted. Funding to implement the acquisition program was to come from either a voter-approved county bond issue or from a Washington state conservation fund.

Finally, after several delays, on September 11, 1978, the King County Council approved the county ordinances necessary to implement a development rights purchase program and submit for voter approval, the issuance of general obligation-bonds for $35 million to fund the program (Dunford 1981). In the November general election, the proposition authorizing the bonds received a "yes" vote from 59.8 percent of the voters, just short of the 60 percent majority needed under Washington state law for validation (Farmlands Study Committee 1979). Convinced that public support necessary for a development rights purchase plan did exist, the County government appointed a new citizen advisory group, Save Our Local Farmlands Committee, to review the defeated ballot measure and determine changes needed for the measure to gain approval. Acting on the suggestions of this committee, the county enacted Ordinance

4341, which provided for a $50 million bond issue to acquire the development rights for the "most important" remaining farm and open space land in the County (King County Council 1979:1). The increased bond issue reflected higher land values as well as a lack of funding assistance from outside sources. Finally, in November 1979, the development rights proposal won endorsement by 63 percent of King County's voters (Dunford 1981).

King County's Purchase Program

King County's transfer of development rights program acquires rights for only the most valuable agricultural land and open space in the county. Approximately 33,000 acres (13,300 ha) of land are eligible under the program (Farmlands Study Committee 1979). All of these lands are situated in the previously designated "King County Agricultural Districts" and encompass identified "Agricultural Lands of County Significance." Further, the eligible lands are divided into three purchase priorities, primarily based on the degree of urban threat. The priority formula was developed so that purchase funds would be spent in the most effective manner. Program organizers were concerned that poorer quality agricultural acreage and less-threatened land might consume large amounts of the funds (Farmlands Study Committee 1979:5). The purchase priorities, along with a complementary bidding system, were designed to insure that the development rights for the highest quality and most endangered agricultural resources would be purchased first and that farmland of lesser value could be purchased with the remaining monies (Dunford 1981:20).

Priority 1 lands, those most threatened by urban development, encompass approximately 7,700 acres (3,100 ha; Figure 15). Included in this group are high-value agricultural resources situated near urban population centers and rapidly suburbanizing areas (Farmlands Study Committee 1979). Priority 2 lands encompass roughly 20,200 acres (8,160 ha) of land, primarily concentrated in two large blocks. These areas are the largest active agricultural areas in King County and consist mainly of dairy farms. Geographically, these two areas are located in portions of the county that are still rural; therefore, the threat of imminent urban development was considered less serious (Farmlands Study Committee 1979). The final category, Priority 3 land, covers all the remaining farmland located within the King County Agricultural Districts designated as Agricultural Lands of County Significance. This includes approximately 5,100 acres (2,060 ha) scattered across King County.

The Selection Process

The purchase of development rights takes place through a series of selection rounds. At least one selection period is scheduled every year until all the authorized funds have been expended or until 1985 (Van Almkerk 1981). In the first two rounds only Priority 1 and 2 lands were eligible, but in subsequent rounds all three prorities could enter the bid process. This scheduling procedure was another attempt by program organizers to acquire property rights for the most endangered farmlands first, as well as to concentrate their acquisition monies in areas where economically viable agricultural operations are located (King County Council 1979:9), maintaining the integrity of established King County Agricultural Districts.

At the beginning of each selection round all eligible property owners are notified and receive an application form. At the close of the application period, the County

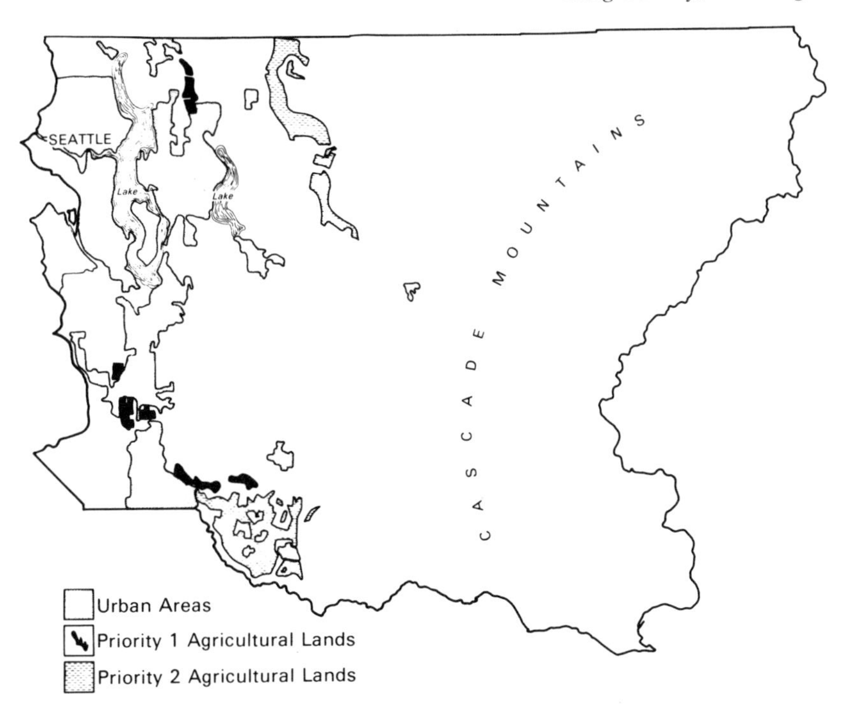

FIGURE 15 AREAS ELIGIBLE FOR PURCHASE OF DEVELOPMENT RIGHTS IN KING COUNTY, WASHINGTON (Compiled by authors from Farmlands Study Committee 1979).

assesses the eligibility of each application and has two appraisals of the property made to determine the value of the development rights. One appraisal determines the fair market value of *fee simple* ownership, while the second is concerned with the fair market value of the agricultural rights only. The difference between the two independent appraisals is the value of the development rights. Both appraisals are provided to the property owners for review.

The ordinance limits payment to no more than the appraised value of the development rights (King County Council 1979:10). Therefore, if the property owner wishes to sell his development rights at the appraised price, the completed applications are turned over to the program administrators and a special seven-member King County Selection Committee made up of agriculturalists and other interested citizens. The Selection Committee and county administrator review all offers and, in the event of inadequate funds in any selection round, establish priorities among the applications. The County Council takes final action on all recommendations and authorizes purchase.

With the completion of the contractual agreements, the county may purchase the development rights in one lump sum or in a series of payments over a period of years. The payment schedule is at the seller's choice. Following the separation of development rights, the property owner is under no obligation to continue farming, the sale simply transfers "the right to use and subdivide land for any and all residential, commercial and industrial purposes and activities. . . ." These development restrictions become part of the property deed as a restrictive covenant in perpetuity.

Program Impacts

In the first 26 months of the King County development rights purchase program, progress toward implementing the policy was slowed by several court challenges (Van Almkerk 1981). Nevertheless, the Priority 1 round of applications was completed. During this first bidding period applications were received on 112 parcels covering 4,003 acres (1,617 ha; Dunford 1981:20), over 60 percent of the eligible land included under Priority 1. The level of landowner participation was surprisingly high, indicating support for the program.

Although the long term effects of King County's purchase program will depend upon continued acceptance by farmers, some potential impacts can be suggested. Land use consequences, particularly a reduction in the supply of land available for urban growth, are not expected to be serious. Consultants hired to examine the economic impacts of the development rights purchase program concluded that the amount of land eligible for the program was insignificant in relation to the total available and suitable land needed for future urban development (Singer Associates 1978). The consultant's report noted that some 130,000 acres (52,520 ha) of vacant land already zoned for residential use is available for new development within the urbanized portion of King County. The Economic Development Council of Puget Sound (1979) also reported there was enough currently zoned land to meet the industrial and warehousing needs of King County for the next 43 years. While the acquisition of development rights will reduce the supply of buildable land, the reduction is not expected to affect either the cost or pattern of future urban growth.

The expected impacts on the King County property tax base are perhaps more significant. The effects will be twofold. First, the value of the development rights purchased by the County will be removed from the tax base. Therefore, the total tax base will be reduced slightly by the program. Second, once the property owner sells his development rights, he can be expected to enroll his land in Washington's differential tax program, the Open Space Tax Act. Assuming large numbers of land owners enroll in the State program, the property tax base could be further reduced. A summary analysis by King County indicates that the financial cost of the development rights purchase program will be an estimated $9.00 per year in increased property taxes for the owner of a $50,000 house over the next 30-year period (Farmlands Study Committee 1979:5).

Perhaps a more serious question is the effect of the purchase program on King County agriculture. Program organizers estimated that the cost of purchasing development rights would average $2723 per acre. Using this figure, the County predicted it would cost $74.2 million to purchase the development rights to all 27,250 acres (11,000 ha) of Priority 1 and 2 land in King County (Farmlands Study Committee 1979: Exhibit 3). Given escalation in land values and delay in implementation, the present $50 million purchase fund will undoubtedly save much less farmland. Whether enough land will be protected to insure the survival of agriculture in King County remains unanswered.

British Columbia: Keeping the Options Open

Public concern over the loss of agricultural land is widespread in Canada, but the environmental and demographic conditions in British Columbia contributed to an early provincial response. In late 1973, the government of British Columbia initiated a series of actions culminating in the passage of Bill 42, the Provincial Land Commission Act, on April 16, 1974. This legislation established comprehensive agricultural land protection policy and an operating framework for its implementation. In subsequent years, British Columbia has been closely watched by other state and provincial governments and has served as the primary model for later Canadian efforts (B.C. Provincial Land Commission 1975).

Background

British Columbia has evolved from and reflects a unique environment. Endowed with a rich resource base, the province has enjoyed a history of almost constant growth and development since joining the confederation in 1856. British Columbia possesses a complex physiography and diverse bioclimatic structure. As a consequence, a robust natural resource-based economy has developed. The forest industry dominates, providing 85,000 direct jobs, but mining, tourism, agriculture, and fishing are also important (British Columbia 1977). Attracted by an expanding economy, mild climate, and the popular mystique surrounding West Coast lifestyle, British Columbia has experienced large scale immigration over the past 30 years. Between 1971 and 1976, the province recorded the highest growth rate in Canada, 12.9 percent.

Spatially, British Columbia's population has always been heavily concentrated in a few urban areas. Today approximately 55 percent of British Columbians live in either the Vancouver or Victoria metropolitan areas, with more than two-thirds of the population crowded into the Lower Fraser Valley (Figure 16). Recent demographic data indicate that population growth continues to be concentrated in the metropolitan regions, but with most new development taking place in the municipal districts on the urban fringe. Clearly, North America's recently dominant urban form — surburban sprawl — has become a part of British Columbia's landscape.

While British Columbia is one of Canada's richest and most diverse provinces, in terms of agriculture it is a have-not. Historically, farming began as an ancillary activity to trading forts and has always been secondary in importance to extractive activities. The most serious obstacle to agricultural development is a severe shortage of agricultural land. With over 90 percent of the province covered by mountains or with elevations over 1,000 meters (Dalichow 1972), British Columbia is aesthetically rich but agriculturally deficient. Estimates of arable land are from 2 to 5 percent of the total land area of the province (Dalichow 1972; Manning and Eddy 1978).

As a result of physiography, British Columbia's agricultural resources may be perceived as "scattered islands nestled between mountain ranges" (Robin 1972). Those pockets of land suitable for food production and situated in proximity to population concentrations have been cultivated since the early 1900s. Intensive cropping and horticultural use has been made of the Fraser Valley, the Okanagan, and southern Vancouver Island. Aggregated, the agricultural lands of the two most important production areas, the Fraser Valley and the Okanagan, encompass only 270,470 acres (109,459 ha) of CLI class 1 to 4 land (B.C. Select Standing Committee on Agriculture 1978).

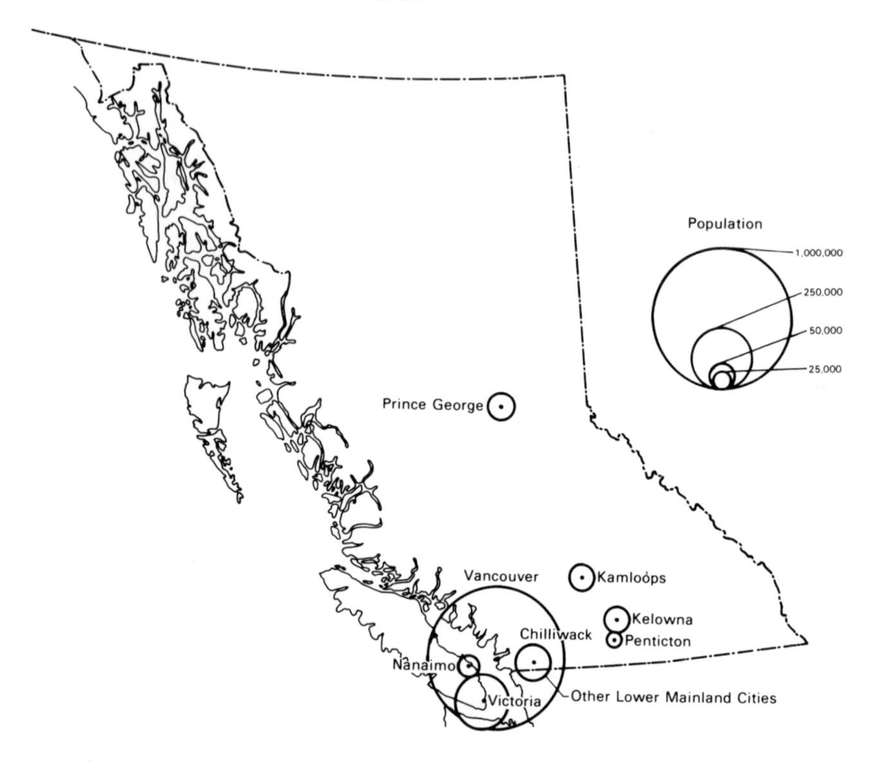

FIGURE 16 BRITISH COLUMBIA POPULATION CONCENTRATIONS. Total 1976 population was 2,184,620.

Not surprisingly, British Columbia is far from self-supporting agriculturally. The province, by necessity, imports agricultural products from other parts of Canada and foreign nations, especially the U.S. and, most recently, Mexico. This dependence has escalated sharply. In 1946, for example, the province required a net import of 3 percent. By 1955, the deficit had risen to 29 percent and amounted to 60 percent in the late 1970s (Baxter 1974; Malzahn 1979). With major dependence on non-Canadian food producers, the province is extremely vulnerable to external political, economic, and climatic conditions which pose the risk of a reduction in food supplies or a sharp increase in food prices.

Complicating the inherent shortage of arable land in British Columbia are the pernicious impacts of urban and population growth. British Columbia's populaton and its most important agricultural production areas are both concentrated in the southern portion of the province. In the competition for space, agricultural activities have been displaced or idled by the higher value urban land uses. One estimate suggested that in the period prior to 1974 over 14,820 acres (6,000 ha) of prime agricultural land were lost annually to urban sprawl in British Columbia (Environment Canada, Lands Directorate 1980:19).

As local attempts to reduce agricultural land conversion made little headway, public attention and support for provincial policies grew stronger. Recognizing the new public mood, three of the four political parties participating in the provincial election campaign of 1972 outlined specific policies to protect agricultural resources. The incumbent Social Credit (Socred) Party campaigned primarily on its past performance, a continuation of local rather than provincial initiatives. The New Democratic Party (NDP), however, proposed a comprehensive ". . . land-zoning program to set aside areas for agricultural production and to prevent such land being subdivided for industrial and residential purposes" (Baxter 1974:8). The weak Liberal and Progressive-Conservative Parties also advocated farmland protection actions. On August 31, 1972, the NDP announced that it would form the new government as a result of winning a majority of seats in the Legislative Assembly. Within three months, the new government moved to implement its promise to protect and encourage agricultural development in the province.

British Columbia's Farmland Protection Program

During its early months in office one of the major tasks of the NDP government was to formulate specifics of its "land-zoning programme" to protect farmland. The strategy evolved was far more sophisticated and wide-ranging than originally expected. The program and its attendent requirements are outlined in Bill 43, The Land Commission Act (later, Agricultural Land Commission Act). As promised, this legislation provided for a wide-ranging strategy to protect and foster agricultural development in British Columbia, while encompassing objectives and policies not attempted in previous North American farmland protection programs. Beyond the preservaton of agricultural land, the powers of the Act included the creation of green belts around cities, development of urban and industrial land banks and reserves, and setting aside land for recreation use (British Columbia 1973).

The Act is unique in its specific attention to the needs of the family farm. This concern for small-scale, family agriculture reflects the philosophical orientation of the NDP and, perhaps, a fear of foreign-owned corporate farm operations moving into British Columbia (Bray 1980). The Act provides a variety of incentives to assist and encourage young people to enter farming and ranching and to aid farming families. Another significant difference between the Act and other North American farmland preservation programs is its attention to the stability of the entire agricultural industry, not simply the protection of land resources. As the B.C. Provincial Land Commission (1977) stated:

> *Legislation and subsequent regulatory zoning is only one aspect of preserving agricultural land in British Columbia. The land resource, while extremely important, is only part of the team. Other partners are the food producers and the farm community infrastructure. The Commission feels strongly that, as well as preserving the agricultural land resource, we must also work to preserve the expertise of the farmer and project the sense of identity, self confidence and vitality of the farm community, if we are to be successful in the long term.*

Consequently, the government introduced and passed several complementary pieces of legislation designed to assist farmers financially. These were the Agricultural Credit Act, which provides for loans and loan guarantees for the purchase of land or improvements, and the Income Assurance Act, which provides income protection for farmers within the Agricultural Land Reserves (Bray 1980). The effect of this corollary legislation is to compensate farmers unable to take advantage of the increased urban-related land values by assuring that reasonable agricultural operations provide a fair return on investment.

The Provincial Land Commission

Structurally, the Agricultural Land Commission Act assigns the administrative and decision-making powers of the program to the provincial government and its agent, the Provincial Land Commission. The Commission is composed of not less than 5 members appointed and serving at the pleasure of the government. The specific powers assigned to the Commission include zoning and regulatory controls over agricultural resources and the Agricultural Land Reserves. The Commission has from its inception, however, sought an expanded activist role, beyond being a zoning administrator. It has defined its niche as ". . . Ombudsman, Advocate, and Catalyst in a variety of situations that related directly or indirectly to the farm community" (B.C. Provincial Land Commission 1978:29). The Commission, in cooperation with agricultural interests, has filed legal briefs with provincial and federal agencies considering actions that influence agriculture and has publicly rebuked government units for failing to support agricultural land preservation goals (B.C. Provincial Land Commission 1976).

The Agricultural Land Reserve

From the outset, the central element in British Columbia's farmland protection program has been the Agricultural Land Reserve (ALR), province-wide agricultural zoning. The Reserve covers 1,862,600 acres (752,490 ha) or 4.9 percent of British Columbia, including both private and Crown-owned land that have the physical capability to grow food (Figure 17). Current agricultural use is not necessary. All land resources with Canada Land Inventory Soil Classes 1 to 4 not required for urban growth during the next 5 years are included in the ALR. Designation as ALR limits future land use to agricultural activities or a variety of uses compatible with the long-term capability of the land to produce agricultural products. The test of whether or not a use is compatible centers around "irreversibility effects (on) the agricultural productivity of the land" (B.C. Provincial Land Commission 1974:5).

From a planning perspective, the ALR may be characterized as an exclusive farm use (EFU) zone. Exclusive farm use zoning has been implemented in only a few other areas on a large scale. A comparison of British Columbia's Agricultural Land Reserve with other forms of farm use zoning reveals a basic structural similarity with significant differences. One of the most critical is the primacy of Land Reserves. Under the language of the Land Commission Act, the designation of Land Reserves supercedes all other land planning and zoning decisions. Local governments or individuals wishing to shift land out of the Reserve or engage in activities that might affect the integrity of ALR must petition the Commission or the Provincial government. The permanence and importance attached to the Land Reserve is unique in planning for privately owned land in North America. A second difference between ALR and other EFU zones surrounds permitted land uses. The legislation and rules of the Land Reserve are far more stringent than the regulations most EFU zones used in other areas. In Oregon, for example, residential development is permitted on large blocks of agricultural land (Furuseth 1980b). This practice is strictly controlled in British Columbia. As a result of rigorous controls, a gradual loss of agricultural production through incremental changes in land use can be minimized.

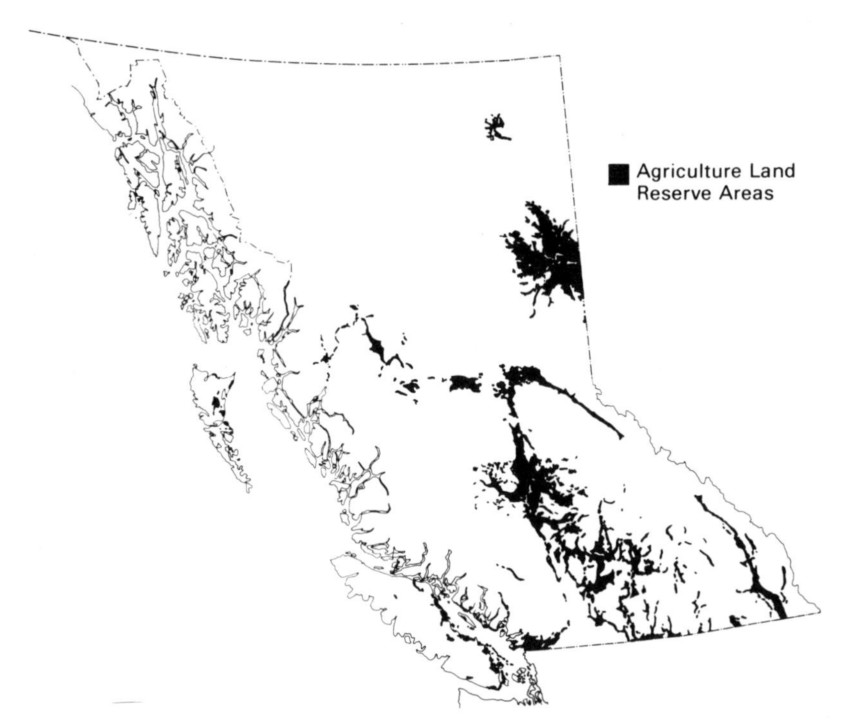

FIGURE 17 BRITISH COLUMBIA'S AGRICULTURAL LAND RESERVE (*After* Environment and Land Use Committee Secretariat 1975).

Program Success

As noted earlier, the ultimate test of whether a public policy is worth the energy and cost of implementation is based on two considerations. Does the policy have the support and backing of the public it is designed to service? Does it succeed in achieving its objectives?

The results of several recent public opinion surveys indicate that the first test of policy success for British Columbia's farmland protection policy is easily met. The 1978 impact study by Manning and Eddy (1978:91) found strong support among landowners and users directly affected by the program. In many cases early opponents of the policy have now become strong proponents. In general, 80 percent of the respondents were favorable toward the provincial program of protecting farmland.

These findings are substantiated by later survey research. Independent public issue surveys conducted by the Federal Government and City of Vancouver in 1978 and 1979 found farmland protection to be a strongly held concern among residents of the Vancouver metropolitan region (Raynor 1980; Vancouver City Planning 1979). According to the survey results prepared for the Federal Ministry of State for Urban Affairs, the protection of farmland from urban development ranked fourth of 26 public concerns. This attitude was reflected in October 1979 and November 1980 when strong public opposition and criticism was directed at the Provincial Cabinet for overturning the recommendations of the Agricultural Land Commission and acting to remove land from the ALR (Malzahn 1979). Through the efforts of envirionmental organizations and *ad hoc* farmland protection groups, attempts to permit development of agricultural resources were publicized. The resulting public protests acted as a deterrent and bolstered the work of the Land Commission.

The second test of policy effectiveness is more difficult to assess. Without data for agricultural land conversion rates before the Act, an analysis of post-legislation conversion activity is flawed. Surprisingly, little quantitative data are available for the period prior to 1973, other than estimates that farmland losses had reached 6,000 ha (14,820 acres) per year.

The only evaluative research of British Columbia's farmland protection policy has been carried out by Pierce (1981a). His analysis of conversion activity within the ALR between 1974 and 1978 found that land transfer out of the ALR is slow, certainly much slower than farmland conversion before the Act. The bulk of land being removed from the Reserves is not in the shadow of major urban metropolitan areas, as was the case before the policy, but rather in rural sections of southern British Columbia and on Vancouver Island. More serious, however, are Pierce's findings relating to the quality of land being taken out of the Reserves. One-third of the removed land was CLI Classes 1-3 (prime land) with the removal rate for prime land increasing during the 5 year study period. While this finding is an ominous trend which should be corrected for long term stability, the conclusions drawn by Pierce are that land withdrawal from the ALR "does not pose a threat to their integrity. The removal of land amounted to a little over one half of one percent of the existing area (of the ALR)" (Pierce 1981a:54). These findings confirm the intuitive observations and conclusions of other Canadian researchers. Manning and Eddy (1978), Bryant and Russwurm (1979), Krueger (1977b), and Malzahn (1979) have all cautiously suggested that British Columbia's farmland protection program has effectively retarded the loss of farmland. Further, all agree that this program is the most promising strategy in Canada.

Based on these findings, it is reasonable to conclude that the Agricultural Land Commission Act promises to become a successful deterrent to the unplanned conversion of agricultural resources. Evidence suggests that the current program can provide workable protection while fostering agricultural development in the province. Clearly, British Columbia has implemented a farmland protection scheme which works, a real achievement in North America.

7

Agricultural Land Resources for the Future

Much of the prosperity of North American society can be traced to the strength and vitality of its agriculture. The continued successful operation of North America's "food producing machine" depends upon society's ability to identify potential long-term problems in agriculture and upon a willingness to address problems with appropriate remedial strategies. We see the urbanization of agricultural land as one such potential long-term problem.

Three research questions have guided our inquiry. What are the basic issues surrounding the urbanization of agricultural land? How has society responded to these issues? And, how effective have these responses been? Answers to these questions lead to a fourth and final query: Where do we go from here?

Perhaps the dominant issue which emerges from our analysis of land conversion data is that the growth and redistribution of population in the United States is consuming more cropland than had previously been estimated and that the reserve of cropland is smaller than originally thought. Consequently, most of this reserve will be needed by the year 2000 to meet anticipated growth in agricultural demand. These observations are, of course, based upon certain assumptions regarding expected gains in agricultural productivity, commodity prices, and changes in demand. If productivity gains keep pace with the growth in demand without serious environmental costs, then there will be little need for additional land to be brought into agricultural production.

The land resource situation is even more critical in Canada because of smaller, less productive reserves and because Canadian urban centers are consuming a greater proportion of prime land than their American counterparts. Given Canadian circumstances, continued increases in cropland productivity will be required to avoid serious constraints on the future productive potential of Canadian farming.

Acceptance of agricultural land protection as a legitimate and preferred action for government has become quite widespread, both geographically and at different levels of government. The basic rationale for societal action is varied. Concern over the adequacy of the land resource base to meet nutritional needs of future generations (the equity question), the indirect impact of urban growth on the local and regional agricultural base (the externality question), and the cost of urban growth (the efficiency question) have all been mentioned or identified by analysts.

Response to these issues by state and provincial governments has varied in terms of timing, scope, design, and, of course, impact. Variation in timing is largely due to the lags in the acceptance of the need to regain land use and taxation powers previously ceded to local governments. Differences in the scope and design of particular programs are linked to the degree of urban pressure on agricultural land, public perception of the farmland problem, and the prevailing political ethos. Finally, the variation in impact has been related to the directness and comprehensiveness of farmland preservation programs.

Three broad classes of farmland preservation approaches were identified in Chapter 5. The first, financial compensation, was most often implemented by use-value taxation, tax credits, and transfer of development rights. Tax based strategies have been judged relatively ineffective in retarding farmland loss owing to insufficient compensation to participants and because participation is voluntary. TDR's are also potentially ineffective since the limited funding earmarked for compensation payments may constrain the geographical extent of farmland protection.

The second major approach, the use of police power, is the traditional means for organizing the spatial spread of the city and minimizing land use conflicts. Whether instituted at the local or provincial (state) level through zoning regulations, subdivision control, or lot size controls, the police power approach has several serious flaws. These include an urban bias, lack of permanence, and very often an incremental operating procedure. Although police powers in Canada are practiced in a comprehensive manner, the failure to attack directly the issue of farmland alienation continues to place farming on the defensive vis-à-vis urban growth pressures.

Finally, we examined comprehensive strategies which fell into two broad categories — voluntary agricultural districts and involuntary exclusive agricultural zoning. The effectiveness of these two approaches is mixed. Evidence to date indicates that insufficient financial incentives combined with weak regulations limit long-term success of agricultural districts in areas undergoing urbanization. The districts appear to be quite successful and popular in rural settings, but fail to perform as designed in farmbelts surrounding urban centers.

The hallmarks of exclusive agricultural zoning are involuntarism, farm income assurance, and a general integrated approach to land use allocation. The exclusive agricultural zoning approach has generally been effective at minimizing the alienation of agricultural land for two very important reasons. First, participation is mandatory and ultimate authority for land use decisions is vested in a commission or some central authority. Second, farmland protection is wedded to initiatives to improve the economic environment for farming.

In addition to minimizing farmland alienation, comprehensive mandatory programs have also reduced haphazard urban development. Aside from these successful features, comprehensive programs enjoy political acceptance, but they are not without criticism. Areas of concern include the degree of centralized control; restrictions on the sale and subdivision of land for non-agricultural purposes; constraints on the expansion of some urban areas; and, perhaps most importantly, questions regarding the need for such programs. The acceptance of mandatory farmland preservation programs will depend to a large degree on the ability of states and provinces to address and resolve these issues.

Arguments used against comprehensive mandatory approaches should not prevent their adoption. Opposition based on excessive centralized control of land use is ill-informed. The Oregon example demonstrates that widespread decision-making powers can be vested with local government, and increased requirements for public involvement can open up the planning process to widespread citizen participation (Furuseth 1980b). Secondly, there are no necessary constitutional roadblocks to actions restricting the sale and subdivision of land for non-agricultural purposes. Rather, compensation appears to be more of a moral or ethical question than a legal one. Opinion is divided over whether the restrictions placed on a change of use for a rural parcel are equivalent to the restrictions placed on an urban parcel under normal zoning controls. The third issue, the economic and social costs of constraints on urban expansion, is more difficult to estimate. This issue is unlikely to be important within the many Canadian and U.S. cities with large quantities of vacant land; where low-density development is very costly to service; or, finally, where a sufficient quality of poor quality land which would be suitable for development. Nevertheless, the long-term cost implications limiting urban expansion require further investigation. If policy is to be equitable for both present and future generations, a full accounting of social costs and benefits will have to be made.

It is the last of the four concerns which is the most difficult to answer unequivocally. Do we need these mandatory programs to prevent negative externalities, to preserve open space and regional agricultural economies, and, above all, to ensure that North America has sufficient land resources to meet its nutritional and economic requirements for the foreseeable future? We have paid particular attention, both conceptually and empirically, to the question of adequacy of the land resource. Our estimates of the impact of urbanization on rural land lead us to the following conclusion: What is critical is not that 4 percent of the U.S. land mass will be urbanized in 20 years' time, but that the equivalent of 10 percent of the potential cropland and half of the potential prime land currently in reserve will be converted to urban uses. That urban growth is consuming a disproportionate share of the cropland base when most, if not all, of the agricultural reserve will be needed in the future points to the need for improvement in the allocation of rural land.

How can we improve the allocation of rural land? Some economists would prefer to leave the allocation of land to the operation of the 'free' market. We argue that farmland protection is not an economic issue. Instead ". . . it is a conservation issue with an economic dimension" (Sampson 1979:96). As a consevation issue, the question of how much land should be consumed today and how much should be left for future generations becomes critical. We recommend that provincial, state, and even local governments adopt the principle outlined in Chapter 1 — a safe minimum standard of conservation for agricultural resources. Such a policy would require governments to define a critical zone or minimum standard for food production which would translate into a corresponding use rate and land use requirement. Far from being a static phenomenon, the critical zone itself would change as we reappraise our "presently foreseeable conditions in technology, wants, and social institutions" (Ciriacy-Wantrup 1965:575).

The application of a safe minimum standard would result in some social costs since urban development would be foregone. However, as in other areas of social policy, such as public health and fire protection, individual costs are small in relation to potential losses to society. Ciriacy-Wantrup (1965:577) argues that "it is impractical to determine the social optimum in the state of public health and safety or national defense because of uncertainty and because of difficulties of evaluating social revenues and costs. On the other hand it is possible to set up standards to avoid serious losses . . ."

Above all else, the safe minimum standard, if applied judiciously as a *goal* of farmland preservation programs, will ensure choice in future land use decisions and, hence, continued flexibility in the economic and social development of North America.

Appendix: Commonly Used Acronyms

The large number of governmental agencies, research reports, and terminology cited in this monograph necessitated a liberal use of acronyms. Listed below is a directory of these abbreviations.

ACRI	Agroclimatic Resource Index (Canada)
ALR	Agricultural Land Reserve (Canada)
CLI	Canada Land Inventory
CMA	Census Metropolitan Areas (Canada)
CNI	Conservation Needs Inventory (U.S.)
EFU	Exclusive Farm Use (Zoning) (U.S.)
LCC	Land Capability Classification System (U.S.)
NALS	National Agricultural Lands Study (U.S.)
NRI	National Resource Inventory (U.S.)
OMB	Ontario Municipal Board (Canada)
PALS	Preservation of Agricultural Lands Society (Canada)
PCS	Potential Cropland Study (U.S.)
RSRI	Regional Science Research Institute (U.S.)
SCC	Soil Capability Classification (Canada)
SCS	Soil Conservation Service (U.S.)
TDR	Transfer of Development Rights (U.S.)

Bibliography

Alaska. 1979. *State Register,* Chapter 70, Natural Resources. Juneau: Alaska State Government.

Barlowe, R. and T. R. Alter. 1976. *Use-Value Assessment of Farm and Open Space Lands.* East Lansing: Michigan State University Agricultural Experiment Station.

Barrows, R. and D. Yanggen. 1978. "The Wisconsin Farmland Preservation Program," *Journal of Soil and Water Conservation* 33:209-212.

Baxter, D. 1974. *The British Columbia Land Commission Act – A Review.* Vancouver: Faculty of Commerce and Business Administration, University of British Columbia, Urban Land Economics Report #8.

Beaubien, C. and R. Tabacnik. 1977. *People and Agricultural Land, Perceptions 4.* Ottawa: Science Council of Canada.

Berry, B. J. L. 1980. "The Urban Problem," pp. 37-59 in A. M. Woodruff (editor), *The Farm and the City: Rivals or Allies?* Englewood Cliffs, NJ: Prentice-Hall.

Berry, B. J. L. and J. D. Kasarda. 1977. *Contemporary Urban Ecology.* New York: MacMillan.

Berry, D. 1978. "Effects of Urbanization on Agricultural Activities," *Growth and Change* 9:2-8.

Berry, D. *et al.* 1976. *The Farmer's Response to Urbanization: A Study of the Middle Atlantic States.* Philadelphia: Regional Science Research Institute.

Beesley, K. B. and L. H. Russwurm (editors). 1981. *The Rural-Urban Fringe: Canadian Perspectives.* Downsview, Ontario: Department of Geography, Atkinson College, York University.

Best, R. H. *et al.* 1974. "The Density-Size Rule," *Urban Studies* 11:201-208.

Blumenfeld, H. 1959. "The Tidal Wave of Metropolitan Expansion," *Journal of the American Institute of Planners* 20:3-14.

Blumenfeld, H. 1965. "The Modern Metropolis," pp. 25-39 in *Scientific American* (editor), *Cities.* New York: Alfred A. Knopf.

Blumenfeld, H. 1967. "Are Land Use Changes Predictable?" pp. 321-326 in P. D. Speirengen (editor), *The Modern Metropolis: Its Origin, Growth Characteristics and Planning.* Montreal: Harvest House.

Bogue, D. J. 1956. *Metropolitan Growth and the Conversion of Land to Nonagricultural Uses.* Oxford, OH: Scripps Foundation.

Bosselman, F. 1976. "Constitutional Issues in Land Use Planning: An Introduction," pp. 65-73 in R. Cowart (editor), *Land Use Planning, Politics and Policy.* Berkeley: University of California Extension Publications.

Bosselman, F. and D. Callies. 1971. *The Quiet Revolution in Land Use Control.* Washington, DC: U.S. Government Printing Office.

Bosselman, F. *et al.* 1973. *The Taking Issue.* Washington, DC: U.S. Government Printing Office.

Bourne, L. S. 1978. "Some Myths of Canadian Urbanization: Reflections on the 1976 Census and Beyond," pp. 124-139 in R. M. Irving (editor), *Readings in Canadian Geography* (3rd edition). Toronto: Holt, Rinehart and Winston.

Boyce, R. 1966. "The Edge of the Metropolis — The Wave Theory Analog Approach," pp. 31-40 in J. V. Minghi (editor), *The Geographer and the Public Environment.* B.C. Geographical Series #7. Vancouver: Tantalus Research.

Bray, C. E. 1980. "Agricultural Land Regulation in Several Canadian Provinces," *Canadian Public Policy* 6:591-604.

British Columbia. 1973. *Land Commission Act,* Chapter 46. Victoria: The Queen's Printer.

British Columbia. 1977. *British Columbia Facts and Figures.* Victoria: The Queen's Printer.

British Columbia. Provincial Land Commission. 1974. *Annual Report April 1, 1973 – March 31, 1974.* Burnaby: Provincial Land Commission.

British Columbia. Provincial Land Commission. 1975. *Annual Report April 1, 1974 – March 31, 1975.* Burnaby: Provincial Land Commission.

British Columbia. Provincial Land Commission. 1976. *Annual Report April 1, 1975 – March 31, 1976.* Burnaby: Provincial Land Commission.

British Columbia. Provincial Land Commission. 1977. *Annual Report Year Ended March 31, 1977.* Burnaby: Provincial Land Commission.

British Columbia. Provincial Land Commission. 1978. *Annual Report Year Ended March 31, 1978.* Burnaby, BC: Provincial Land Commission.

British Columbia. Select Standing Committee on Agriculture. 1978. *Inventory of Agricultural Land Reserves in British Columbia. Phase I.* Victoria, BC: Legislative Assembly.

Brown, H. J. *et al.* 1981. "Land Markets at the Urban Fringe," *Journal of the American Planning Association* 47:131-144.

Bryant, C. R. 1973. "The Anticipation of Urban Expansion," *Geographia Polonica* 28:93-115.

Bryant, C. R. 1976. *Farm-Generated Determinants of Land Use Changes in the Rural-Urban Fringe in Canada, 1961-1975.* Lands Directorate, Environment Canada. Ottawa: Supply and Services Canada.

Bryant, C. R. and L. R. Russwurm. 1979. "The Impact of Non-Farm Development on Agriculture: A Synthesis," *Plan Canada* 19:122-139.

Bryant, C. R. and L. H. Russwurm. 1981. "Agriculture in the Urban Field: Canada, 1941 to 1971" pp. 34-52 in Beesley and Russwurm (1981).

Bryant, W. R. 1975. *Farmland Preservation Alternatives in Semi-Suburban Areas.* Ithaca, NY: Cornell University, Department of Agricultural Economics, A. E. Extension Report No. 75-1.

Canada, Department of Regional Economic Expansion. 1965. *The Canada Land Inventory Soil Capability Classification for Agriculture.* Ottawa: The Queen's Printer.

Chapman, L. J. and D. M. Brown. 1966. *The Climates of Canada for Agriculture.* Canada Land Inventory Report No. 3. Ottawa: Queen's Printer.

Ciriacy-Wantrup, S. V. 1965. "A Safe Minimum Standard as an Objective of Conservation Policy," pp. 575-584 in I. Burton and R. W. Kates (editors), *Readings in Resource Management.* Chicago: University of Chicago Press.

Clawson, M. 1971. *Suburban Land Conversion in the United States.* Baltimore: John Hopkins Press.

Committee for the Preservation of Agricultural Land. 1973. *Final Report to the Secretary of Agriculture.* Annapolis: Maryland Department of Agriculture.

Conklin, H. E. (editor). 1980. *Preserving Agriculture in an Urban Region.* Ithaca, N.Y.: Cornell University, Agricultural Experiment Station, Bulletin 86.

Conroy, R. T. 1978. *Preserving Prime Agricultural Land in the United States.* Plattsburgh, NY: Institute for Man and Environment, State University of New York at Plattsburgh, Regional Studies Report No. 14.

Conservation Foundation. 1981. *Growing Pains: Rural America in the 1980s.* (Movie). Washington, DC: The Conservation Foundation.

Coughlin, R. 1979. "Agricultural Land Conversion in the Urban Fringe," pp. 29-48 in Schnepf (1979).

Coughlin, R. E. *et al.* 1977. *Saving the Garden: The Preservation of Farmland and Other Environmentally Valuable Land.* Philadelphia: Regional Science Research Institute.

Coughlin, R. E. *et al.* 1981. *The Protection of Farmland: A Reference Guidebook for State and Local Governments.* Washington, D.C.: U.S. Government Printing Office.

Crerar, A. D. 1960. "The Loss of Farmland in the Growth of Metropolitan Regions of Canada," pp. 181-195 in *Resources for Tomorrow*, Supplementary Volume. Ottawa: The Queen's Printer.

Crosson, P. 1977. "Demands for Food and Fiber: Implications for Land Use in the U.S.," pp. 49-61 in *Land Use: Tough Choices in Today's World.* Ankeny, IA: Soil Conservation Society of America.

Crosson, P. 1979. "Agricultural Land Use: A Technological and Energy Perspective," pp. 99-112 in Schnepf (1979).

Dalichow, F. 1972. *Agricultural Geography of British Columbia.* Vancouver: Versatile Publishing.

Davies, R. and J. Belden. 1979. "A Survey of State Programs to Preserve Farmland." Paper presented at the National Conference of State Legislators. Washington, DC.

Davis, K. 1965. "The Urbanization of the Human Population," pp. 3-24 in *Scientific American* (editor), *Cities.* New York: Alfred A. Knopf.

Dideriksen, R. I. *et al.* 1977. *Potential Cropland Study.* Statistical Bulletin 578. Washington, DC: U.S. Government Printing Office.

Dideriksen, R. I. *et al.* 1979. "Trends in Agricultural Land Use," pp. 13-28 in Schnepf (1979).

Dideriksen, R. I. and R. N. Sampson. 1976. "Important Farmlands: A National View," *Journal of Soil and Water Conservation* 31:195-197.

Dill, H. W. and R. C. Otte. 1971. *Urbanization of Land in the Northeastern United States.* Economic Research Service Report 485. Washington, DC: U.S. Government Printing Office.

Drewett, R. 1971. "Land Values and Urban Growth," pp. 335-357 in M. Chisholm *et al.* (editors), *Regional Forecasting.* London: Archon Books.

Dunford, R. W. 1981. "Saving Farmland, The King County Program," *Journal of Soil and Water Conservation* 36:19-22.

Dunford, R. W. 1982. "The Evolution of Federal Farmland Protection Policy," *Journal of Soil and Water Conservation* 37:133-136.

Economic Development Council of Puget Sound. 1979. *A Summary Report on Industrial Land Absorption in King County.* Seattle: Economic Development Council of Puget Sound.

Emanuel, M. S. 1977. "TDR: Rural Town of Eden Uses TDR to Save Agricultural Land," *Practicing Planner* 7:15-20.

Environment and Land Use Committee Secretariat. 1975. *Agricultural Land Reserves of British Columbia* (map). Victoria, BC: Province of British Columbia.

Environment Canada. 1972. *Soil Capability Classification for Agriculture.* Canada Land Inventory Report No. 2. Ottawa: Supply and Services Canada.

Environment Canada. Lands Directorate. 1980. *Land Use in Canada, the Report of the Interdepartmental Task Force on Land-Use Policy.* Ottawa: Supply and Services Canada.

Esseks, J. D. 1978. "The Politics of Farmland Preservation," pp. 199-216 in D. F. Hadwinger and W. P. Browne (editors), *The New Politics of Food.* Lexington, MA: Lexington Books, D.C. Heath.

Farmlands Study Committee. 1979. *Saving Farmlands and Open Space.* Seattle: King County Office of Agriculture.

Foster, J. 1976. *Changes in Agricultural Land Use in Massachusetts, 1951-1971.* Amherst: Massachusetts Agricultural Experiment Station.

Frankena, M. W. and D. T. Scheffman. 1980. *Economic Analysis of Provincial Land Use Policies in Ontario.* Toronto: University of Toronto Press.

Furuseth, O. J. 1980a. "If We Are Really Serious About Protecting Agricultural Land in North Carolina . . ." *Carolina Planning* 6:40-51.

Furuseth, O. J. 1980b. "The Oregon Agricultural Protection Program: A Review and Assessment," *Natural Resources Journal* 20:603-614.

Furuseth, O. J. 1981. "State Farmland Protection Policies: Diffusion of a Public Policy Innovation." Paper presented at the Annual Meeting of the Southeast Division, Association of American Geographers, Atlanta.

Furuseth, O. J. and J. T. Pierce. 1982. "A Comparative Analysis of Farmland Preservation Programmes in North America," *Canadian Geographer* 26:191-206.

Gardner, B. D. 1977. "The Economics of Agricultural Land Preservation," *Journal of Agricultural Economics* 59:1027-36.

Gayler, H. J. 1982. "The Problems of Adjusting to Slow Growth in the Niagara Region of Ontario," *Canadian Geographer* 26:165-171.

Geno, B. J. and L. M. Geno. 1976. *Food Production in the Canadian Environment,* Perceptions 3, Study on Population, Technology and Resources. Ottawa: Science Council of Canada.

Geier, K. E. 1980. "Agricultural Districts and Zoning: A State-Local Approach to a National Problem," *Ecology Law Review* 8:655-696.

Gertler, L. O. 1968. *The Niagara Escarpment Study: Fruit Belt Report.* Toronto: Department of Treasury and Economics.

Gertler, L. O. and J. Hind-Smith. 1961. "The Impact of Urban Growth on Agricultural Land: A Pilot Study," pp. 155-180 in *Resources for Tomorrow,* Supplementary Volume. Ottawa: The Queen's Printer.

Gierman, D. M. 1977. *Rural to Urban Land Conversion.* Occasional Paper No. 16, Lands Directorate, Environment Canada. Ottawa: Supply and Services Canada.

Gloudemans, R. J. 1974. *Use-Value Farmlands Assessment – Theory, Practice, and Impact.* Chicago: International Association of Assessing Officers.

Goodwin, R. K. and W. B. Shepard. 1974. *State Land Use Policies; Winners and Losers.* Man's Activities as Related to Environmental Quality — A Series, No. 5. Corvallis: Oregon State University.

Gottman, J. 1961. *Megalopolis.* New York: Twentieth Century Fund.

Gustafson, G. and L. Wallace. 1975. "Differential Assessment as Land Use Policy: The California Case," *Journal of the American Planning Association* 41:379-389.

Harriss, C. L. 1980. "Free Market Allocation of Land Resources (What the Free Market Can and Cannot Do in Land Policy)," pp. 123-143 in A. M. Woodruff (editor), *The Farm and the City: Rivals or Allies?* Englewood Cliffs, NJ: Prentice-Hall.

Hart, J. F. 1968. "Loss and Abandonment of Cleared Farmland in the Eastern United States," *Annals,* Association of American Geographers 58:417-440.

Hart, J. F. 1976. "Urban Encroachment on Rural Areas," *Geographical Review* 66:3-17.

Heady, E. O. 1976. "The Agriculture of the U.S." pp. 77-86 in *Scientific American* (editor), *Food and Agriculture.* San Francisco: Freeman.

Hoffman, D. W. 1976. "Soil Capability Analysis and Land Resource Development in Canada," pp. 140-167 in G. R. McBoyle and E. Sommerville (editors), *Canada's Natural Environment: Essays in Applied Geography.* Toronto: Methuen.

Holcomb, H. B. and R. A. Beauregard. 1981. *Revitalizing Cities.* Washington, DC: Association of American Geographers, Resource Publications in Geography.

Hoover, F. M. 1968. "The Evolving Form and Organization of the Metropolis," pp. 237-284 in H. S. Perloff and L. Wingo, Jr. (editors), *Issues in Urban Economics.* Baltimore: Johns Hopkins.

Hudson, J. 1973. "Density and Pattern in Suburban Fringes," *Annals,* Association of American Geographers 63:28-39.

Irving, R. M. (editor). 1957. *Factors Affecting Land Use in a Selected Area in Southern Ontario – A Land Use and Geographic Survey of North Township in Lincoln County, Ontario.* Toronto: Ontario Depatment of Agriculture.

Jackson, J. N. 1982. "The Niagara Fruit Belt: The Ontario Municipal Board Decision of 1981," *Canadian Geographer* 26:172-176.

Kaiser, E. J. and S. F. Weiss. 1970. "Public Policy and Residential Development Process," *Journal of the American Institute of Planners* 36:30-37.

Keene, J. C. *et al.* 1976. *Untaxing Open Space.* Washington, DC: U.S. Government Printing Office.

King County. 1977. *King County Agricultural Protection Program, Background and Effects of Ordinance 3064.* Seattle: King County Council.

King County. Office of Agriculture. 1977. *Proposed Agricultural Land Preservation Program.* Seattle: King County Council.

King County Council. 1977. *Ordinance No. 3064.* Seattle: King County Government.

King County Council. 1979. *Ordinance No. 4341.* Seattle: King County Government.

Knebel, J. A. 1976. *Secretary's Memorandum No. 1827, Supplement 1, "Statement on Prime Farmland, Range, and Forest Land,"* Washington, DC: U.S. Department of Agriculture.

Krueger, R. R. 1977a. "The Destruction of a Unique Renewable Resource: The Case of the Niagara Fruit Belt," pp. 132-148 in Krueger and Mitchell (1977).

Krueger, R. R. 1977b. "The Preservation of Agricultural Land in Canada," pp. 119-131 in Krueger and Mitchell (1977).

Krueger, R.R. 1978. "Urbanization of the Niagara Fruit Belt," *Canadian Geographer* 22:179-194.

Krueger, R. R. 1980. "The Struggle to Preserve Specialty Crop Land in the Rural-Urban Fringe of the Niagara Peninsula of Ontario," (unpublished manuscript).

Krueger, R. R. 1981a. "The Struggle to Preserve Specialty Crops," *Kitchener-Waterloo Record,* Dec. 29, 1981.

Krueger, R. R. 1981b. "The Niagara Fruit Belt Controversy Update 1981," *Monograph* 32:9-11.

Krueger, R. R. and B. Mitchell (editors). 1977. *Managing Canada's Renewable Resources.* Toronto: Methuen.

Lapping, M. B. 1980. "Agricultural Land Retention: Responses, American and Foreign," pp. 144-178 in A. M. Woodruff (editor), *The Farm and the City, Rivals or Allies?* Englewood Cliffs, NJ: Prentice Hall.

Latham, R. F. and M. H. Yeates. 1970. "Population Density Growth in Metropolitan Toronto," *Geographical Analysis* 2:177-185.

Lee, L. K. 1978. *A Perspective on Cropland Availability.* ESCS Report No. 406. Washington, DC: U.S. Government Printing Office.

Malzahn, M. 1979. "B.C.'s Green Acres: A Look at the Future of Farmland in B.C.," *Urban Reader* 7:14-19.

Manning, E. W. and S. S. Eddy. 1978. *The Agricultural Land Reserves of British Columbia: An Impact Analysis.* Land Use in Canada Series No. 13, Lands Directorate, Environment Canada. Ottawa: Supply and Services Canada.

Martin L. R. G. 1975. *Land Use Dynamics on the Toronto Urban Fringe.* Ottawa: Environment Canada.

Maryland Agricultural Code. 1974. Sections 2-501 — 2-508 (Supplement 1978).

Maryland Agricultural Code. 1978. Section 2-509 (Supplement 1978).

Maryland Agricultural Land Preservation Foundation. 1980. *Report to the Maryland General Assembly.* Annapolis: Maryland Department of Agriculture.

Maryland Agricultural Land Preservation Foundation. 1981. *Program Summary.* Annapolis: Maryland Department of Agriculture.

Maryland Cooperative Extension Service. 1977. *Maryland Agricultural Land Preservation Foundation: A Summary.* College Park: University of Maryland Cooperative Extension Service.

Maryland Cooperative Extension Service. 1978. *Maryland Agricultural Economics.* No. 26. College Park: University of Maryland Cooperative Extension Service.

Mayo, J. 1966. *Niagara Region Local Government Review: Report of the Commission.* Toronto: Ontario Department of Municipal Affairs.

McCuaig, J. D. and H. J. Vincent. 1980. *Assessment Procedures in Canada and Their Use in Agricultural Land Preservation.* Lands Directorate, Environment Canada, Working Paper No. 7. Ottawa: Supply and Services Canada.

McInerney, J. P. 1981. "Natural Resource Economics: The Basic Analytical Principles," pp. 30-59 in J. A. Butlin (editor), *The Economics of Environmental and Natural Resources Policy.* Boulder: Westview Press.

Metropolitan Dade County Planning Department. 1980. *Dade County Agricultural Land Use Project.* Miami: Metropolitan Dade County Planning Department.

Mills, D. 1973. "Suburban and Exurban Growth," pp. 51-102 in *The Spread of Cities.* Milton Keynes, England: Open University Press.

Mundie, R. 1980. "Farmland Preservation Programs: Evaluating Effectiveness," Paper presented at the American Planning Association Convention, San Francisco.

Musselman, A. R. 1981. Executive Director, Maryland Agricultural Land Preservation Foundation, *private communication,* October 30, 1981.

Myrdal, G. 1957. *Economic Theory and Underdeveloped Regions.* London: Duckworth Press.

National Agricultural Lands Study. 1980. *Agricultural Land Data Sheet.* Interim Report 2. Washington, DC: U.S. Government Printing Office.

National Agricultural Lands Study. 1981a. *An Inventory of State and Local Programs to Protect Farmland.* Washington, DC: U.S. Government Printing Office.

National Agricultural Lands Study, 1981b. *Case Studies on State and Local Programs to Protect Farmland.* Washington, DC: U.S. Government Printing Office.

National Agricultural Lands Study. 1981c. *Final Report.* Washington, DC: U.S. Government Printing Office.

Newling, B. E. 1969. "The Spatial Variation of Urban Population Densities," *Geographical Review* 59:242-252.

Nielsen, C. A. 1979. Preservation of Maryland Farmland: A Current Assessment," *The University of Baltimore Law Review* 3:429-460.

Nowland, J. L. and G. D. V. Williams. *No Date.* "Areas in Agricultural Land Resource Classes, Rated by Soil and Climate," p. 46 in Simpson-Lewis *et al.* (1979).

Ontario. 1970. *Design for Development: The Toronto-Centred Region.* Toronto: The Queen's Printer.

Ontario. 1974. *Central Ontario Lakeshore Urban Complex.* Toronto: The Queen's Printer.

Ontario. Ministry of Agriculture and Food. 1976. *A Strategy for Ontario Farmland.* Toronto: The Queen's Printer.

Ontario. Ministry of Agriculture and Food. 1977. *Green Paper on Planning for Agriculture: Food Land Guidelines.* Toronto: The Queen's Printer.

Ontario. Ministry of Housing. *No Date.* *A Guide to the Planning Act.* Toronto: The Queen's Printer.

People for Open Space. 1980. *Endangered Harvest, The Future of Bay Area Farmland.* San Francisco: People for Open Space.

Plaut, T. R. 1980. "Urban Expansion and the Loss of Farmland in the United States: Implications for the Future," *American Journal of Agricultural Economics* 62:539-542.

Pierce, J. T. 1979. "Land Conversion and Urban Growth: Canada 1966-1971," *Tijdschrift voor Economische en Sociale Geografie* 70:333-338.

Pierce, J. T. 1981a. "The B.C. Agricultural Land Commission: A Review and Evaluation," *Plan Canada* 21:48-56.

Pierce, J. T. 1981b. "Conversion of Rural Land to Urban: A Canadian Profile," *Professional Geographer* 33:163-173.

Pierce, J. T. and O. J. Furuseth. 1982. "Farmland Protection Planning in British Columbia," *GeoJournal* 6(6); *forthcoming.*

Pizor, P. J. *et al.* 1979. *A Transfer of Development Rights Sampler: A Collection of TDR Ordinances from Municipalities in Eight States.* New Brunswick, NJ: Rutgers University Experiment Station.

***Practicing Planner.* 1977.** "Transfer of Development Rights," *Practicing Planner* 6:10-14.

Preston, R. E. and L. H. Russwurm. 1977. "The Developing Canadian Urban Pattern: An Analysis of Population Change, 1971-1976," pp. 1-30 in R. E. Preston and L. H. Russwurm (editors), *Essays on Canadian Urbanization.* Waterloo: Department of Geography, University of Waterloo.

Pryde, P. R. 1982. "Is There Any Hope for Agriculture in California's Rapid Growth Areas?," *GeoJournal* 6(6): *forthcoming.*

Rawson, M. 1977. "Letter to the Department of City Planning, Faculty of Agriculture, University of Manitoba," quoted on p. 68 in Beaubien and Tabacnik (1977).

Raynor I. 1980. *Canadian Public Priorites, The Canadian Mortgage and Housing Corporation Survey.* Ottawa: Supply and Services Canada.

Real Estate Research Corporation of Chicago. 1974. *The Costs of Sprawl, Executive Summary.* Washington, DC: U.S. Government Printing Office.

Robin, M. 1972. *The Rush for Spoils.* Toronto: McCelland and Stewart.

Sampson, R. N. 1979. "The Ethical Dimension of Farmland Protection," pp. 89-98 in Schnepf (1979).

Schiff, S. D. 1979. "Saving Farmland: The Maryland Program," *Journal of Soil and Conservation* 34:204-210.

Schnepf, M. editor. 1979. *Farmland, Food and the Future.* Ankeny, IA: Soil Conservation Society of America.

Schumacher, E. F. 1979. *Good Work.* New York: Harper and Row.

Simpson-Lewis, W., *et al.* 1979. *Canada's Special Resource Lands: A National Perspective of Selected Land Uses.* Map Folio No. 4, Lands Directorate, Environment Canada. Ottawa: Supply and Services Canada.

Sinclair, R. 1967. "Von Thunen and Urban Sprawl," *Annals,* Association of American Geographers 57:72-87.

Singer, John M., Associates, Inc. 1978. *Economic Impacts of Agricultural Land Preservation in King County: Development Rights Purchase.* Seattle: King County Office of Agriculture.

Skolds, M. C. and J. B. Penn. 1977. "Production Potentials in U.S. Agriculture," pp. 77-102 in U.S. Department of Agriculture, *Looking Forward: Research Issues Facing Agriculture and Rural America.* Washington, DC: U.S. Government Printing Office.

Smit, B. and C. Conklin. 1981. "Future Urban Growth and Agricultural Land: Alternatives for Ontario," *Ontario Geography* 18:47-55.

Sobetzer, J. G. 1979. "American Land and Law," pp. 213-218 in R. N. Andrews (editor), *Land in America.* Lexington, MA: D. C. Heath.

Statistics Canada. 1977. *Census – Farms by Size, Area and Use of Land.* Catalogue No. 96-854. Ottawa: Supply and Services Canada.

Stiglitz, J. E. 1979. "A Neoclassical Analysis of the Economics of Natural Resources," pp. 36-67 in K. V. Smith (editor), *Scarcity and Growth Reconsidered.* Washington, DC: Resources for the Future.

Thom, R. 1975. *Structural Stability and Morphogenesis* (Translated by D. Fowler). Reading, UK: Benjamin/Cummings.

Toner, W. 1978. *Saving Farms and Farmlands: A Community Guide.* Planning Advisory Service Report No. 333. Chicago, IL: American Society of Planning Officials.

U. S. Environmental Protection Agency. 1978a. *Background Paper in Support of an EPA Policy to Protect Environmentally Significant Agricultural Lands.* Washington, DC: U.S. Environmental Protection Agency.

U. S. Environmental Protection Agency. 1978b. *EPA Policy to Protect Environmentally Signficant Agricltural Lands.* Washington, DC: U. S. Environmental Protection Agency.

U. S. Soil Conservation Service. 1966. *Land Capability Classification.* Agriculture Handbook 210. Washington, DC.: U.S. Government Printing Office.

U. S. Soil Conservation Service. 1971. *Basic Statistics of the National Inventory of Soil and Water Conservation Needs.* Statistical Bulletin 461, Washington, DC: U.S. Government Printing Office.

U. S. Soil Conservation Service. 1975. *Background Paper: Prime, Unique, and Other Important Farmlands.* Washington, DC: U. S. Department of Agriculture.

U. S. Soil Conservation Service. 1977. *Potential Cropland Study.* Statistical Bulletin No. 578. Washington, DC: U.S. Government Printing Office.

U. S. Soil Conservation Service. 1979. *National Summaries of the 1977 National Resource Inventories.* Washington, DC: U. S. Government Printing Office.

Van Almkerk, P. 1981. Director, King County Office of Agriculture, *private communication,* October 15, 1981.

Vancouver City Planning. 1979. "Goals for Vancouver Program," *Quarterly Review* 6: 4-6.

Vining, D. R. T. *et al.* 1977. "Urban Encroachment on Prime Agricultural Land in the United States," *International Regional Science Review* 2:143-156.

Vogeler, I. 1978. "Effectiveness of Differential Assessment of Farmland in Metropolitan Chicago," *Geographical Survey* 7:23-32.

Walker, G. 1981. "Farmers in Urban Shadow: A Behavioral Profile" pp. 185-198 in Beesley and Russwurm (1981).

Warren, C. L. and P. C. Rump. 1981. *Urbanization of Rural Lands in Canada: 1966-1971 and 1971-1976.* Lands Directorate, Environment Canada, Land Use in Canada Series No. 20. Ottawa: Supply and Services Canada.

Williams, G. D. V. 1973. "Urban Expansion and Canadian Agro-climatic Resource Problem," *Greenhouse-Garden-Grass* 12:15-26.

Williams, G. D. V. 1974. "Physical Frontiers of Crops: The Example for Growing Barley to Maturity in Canada," pp. 108-127 in R. G. Ironside *et al.* (editors), *Frontier Settlement.* Edmonton: Department of Geography, University of Alberta.

Williams, G. D. V. 1975. *An Agroclimatic Resource Index for Canada and Its Use in Describing Agricultural Land Losses* (unpublished). Ottawa: Agriculture Canada Report to Science Council of Canada.

Williams, G. D. V., *et al.* 1978. "The Spatial Association of Agroclimatic Resources and Urban Population in Canada," pp. 165-179 in R. Irving (editor), *Readings in Canadian Geography* (3rd edition). Toronto: Holt, Rinehart and Winston Ltd.

Winsborough, H. H. 1962. "City Growth and City Structure," pp. 239-255 in W. H. Leaky *et al.* (editors), *Urban Economics.* New York: Free Press.

Wood, D. 1977. "Land Use Regulation. Is Compensation Payable?" pp. 70-72 in Beaubien and Tabacnik (1977).

Woodruff, A. M. 1980. "City Land and Farmland," pp. 11-38, in A. M. Woodruff (editor), *The Farm and the City: Rivals or Allies?* Englewood-Cliffs, NJ: Prentice Hall.

Zeeman, E. C. 1976, "Catastrophe Theory," *Scientific American* 234:65-83.

Zeimetz, K. A. *et al.* 1976. *Dynamics of Land Use in Fast Growth Areas.* U.S. Department of Agriculture, Economic Research Service Report 325. Washington, DC: U.S. Government Printing Office.